OPTICAL COMMUNICATION TECHNOLOGY

Edited by **Pedro Pinho**

Optical Communication Technology
http://dx.doi.org/10.5772/67385
Edited by Pedro Pinho

Contributors

Boris I. Lembrikov, Yossef Ben-Ezra, Edward M. D. Fisher, Waleed Tariq Sethi, Hamsakutty Vettikalladi, Habib Fathallah, Mohamed Himdi, Benjamin Dingel, Nicholas Madamopoulos, Andru Prescod, Ali Shahpari, Isiaka Alimi, Artur Neves E Sousa, Ricardo Ferreira, Paulo Monteiro, António Teixeira, Pedro Pinho

Published by InTech

Janeza Trdine 9, 51000 Rijeka, Croatia

Publishing Process Manager Dajana Pemac
Technical Editor SPi Global
Cover InTech Design team

First published July, 2017
Printed in Croatia
Legal deposit, Croatia: National and University Library in Zagreb

Additional hard copies can be obtained from orders@intechopen.com

Optical Communication Technology, Edited by Pedro Pinho
p. cm.
Print ISBN 978-953-51-3417-6
Online ISBN 978-953-51-3418-3

For EU product safety concerns:
IN TECH d.o.o., Prolaz Marije Krucifikse Kozulić 3, 51000 Rijeka, Croatia,
info@intechopen.com or visit our website at intechopen.com.

Meet the editor

Pedro Pinho was born in Vale de Cambra, Portugal, in 1974. He received his Licenciado and master's degrees in Electrical and Telecommunications Engineering and his PhD degree in Electrical Engineering from the University of Aveiro in 1997, 2000, and 2004, respectively. He is currently an assistant professor in Electronics, Telecommunications and Computers Engineering Department in Instituto Superior de Engenharia de Lisboa (ISEL/IPL) and a senior member of the research staff in Instituto de Telecomunicações (IT), Aveiro, Portugal. Dr. Pinho is also a senior member of the IEEE, serves as Technical Program Committee in several conferences, and is a reviewer of several IEEE journals. He has authored or coauthored 1 book, 10 book chapters, and more than 100 papers for conferences and international journals. He participated as a principal investigator or coordinator in projects with scientific and/or industry focus, both at a national and international level. To date, he has led and leads five PhD degree students and 42 MSc degree students. His current research interest is in optical systems and antennas.

Contents

Preface

To Carla Lourenço, Íris Pinho, and Petra Pinho

During the past few decades, information technologies have been evolving at a tremendous rate, causing profound changes to our world and our ways of life. The annual IP traffic is now measured in exabyte. Concepts of optical fibers, optical networking science, new optical technology, and engineering have been playing an increasingly crucial role in the success of this telecommunication revolution. Optical links provide large bandwidth, and optical fibers are the only medium that can meet modern society's needs of massive amounts of data over long distances, to support all kinds of data-intensive applications such as video, voice, and data services. Considering all this, one can say there are endless possibilities for optical communications.

This book aims to introduce and treat a series of advanced and emerging topics in the field of optical communications. Namely, one of the main subjects of this book is the challenges and the new opportunities that will appear in optical wireless communication. Moreover, there are also contributions related to optical hardware, namely, lasers, modulators, semiconductor optical amplifiers, and optical receivers, discussed in the light of how they contribute to improve the photonic links.

The first chapter is a road map to the optical communication for the next years, the challenges, and the expected evolution. The second one is dedicated to optical wireless communication, where the challenges and opportunities of this technology are presented.

The third chapter presents the design and simulated results of two new optical nanoantenna working at center frequency of 193.5 THz (1550 nm wavelength). Since most of the antennas are metallic based with noble metals as radiators, the conducting losses are huge. One solution presented in this chapter is to integrate the nanoantennas with dielectric resonator.

The fourth chapter is about quantum dot (QD) laser devices. In particular, quantum dot-in-a-well (QDWELL) lasers are characterized by an extremely low threshold current density and high modulation frequency. The chapter presents the theoretically nonlinear optical phenomena in QDWELL lasers and amplifiers under optical injection. It shows that the synchronization of the carrier dynamics in QD and quantum well (QW) caused by the optical injection improves the QDWELL laser performance and enhances the relaxation oscillation frequency. As a result, the QDWELL laser performance in the analogous optical link is significantly improved.

Electro-optic modulators are a critical part of optical communication chains. An essential requirement is that they have a linear response, with a high dynamic range, to high input RF (radio frequency) power of multiple tones. In the fifth chapter, it is shown that interferometric modulator with phase-modulating and cavity-modulating components (IMPACC) can operate with highly linear response with a peak spurious-free dynamic range (SFDR) value of 132 dB.

The last chapter covers the specifics of both optical detection using the internal photoelectric effect and how avalanche multiplication can be extended to runaway avalanche producing full logic-level voltage pulses with single-photon sensitivity. The application, visible light communications, highlights that such receivers require further study before all noise sources and all circuit topologies are understood fully with respect to the higher-level communication performance metrics.

The editor would like to express his gratitude to the publisher for the assigned time, invaluable experience, efforts, and staff who successfully contributed toward enriching the final overall quality of the book.

Pedro Pinho
Instituto Superior de Engenharia de Lisboa (ISEL), Lisbon, Portugal
Instituto de Telecomunicações (IT), Aveiro, Portugal

Chapter 1

Introductory Chapter: The Path to the Future

Pedro Pinho

Additional information is available at the end of the chapter

http://dx.doi.org/10.5772/intechopen.70216

1. Introduction

The idea to use light for communication purposes was envisioned many years ago. However, the origins of optical communications trace back to 1966 when Charles Kao suggested the possibility that low-loss optical fibers could be competitive with coaxial cable and metal waveguides for telecommunication applications. However, it was not until 1970 when Corning Glass Works announced an optical fiber loss less than the benchmark level of 10 dB/km that commercial applications began to be realized. Quickly, the purifying process was refined, and in the late 1980s, optical fibers with an attenuation of 0.2 dB/km were manufactured and commercialized. Surely, our society was shaped by the concepts developed by Charles Kao in the 1960s. In recognition of his pioneering work, in 2009, Charles Kao was awarded the Nobel Prize in Physics along with Willard S. Boyle and George E. Smith.

What followed from 1970 to the first decade of twenty-first century was an extraordinary race that has led to an enormous increase in both the maximum transmission capacities and the transmission distances, and, therefore, to the contents delivered. Nowadays, in addition to the voice, data, and video, it is common talk references to global internet video, virtual reality, augmented reality, and gaming highlights. All of these applications will be a reality in the next years and will be supported by an optimized optical network. However, this race for ever better performance does not stop here. In the upcoming years, developments in several areas of optical communication from hardware, software (algorithms), signal processing to networks and emerging technologies [2] will be the target of much attention from academia and industry.

2. Challenges and Opportunities for the next years

The optical hardware is related to the need for low losses, low dispersion, high speed, and high capacity. So, in this topic, it is possible to cover the following areas:

- New optical fibers for next generation optical networks;
- Optical amplification and regeneration;
- Spatial multiplexing;
- Coherent transceivers.

In relation to the software and signal processing, the main objective is to improve encoding of digital data onto lightwaves and to recover the information reliably at the receiver with the maximum efficiency. Examples of some topics related to this thematic are as follows:

- New modulation formats;
- Digital signal processing;
- Optical signal processing;
- Nonlinear channel modeling and mitigation and forward error correction (FEC).

Then follows the block of networks, where the main objective is to improve the number of point-to-point links between two extremes in different types of networks: core, metropolitan, access, and data center networks. Topics under development in this area are as follows:

- Long-haul networks;
- Access networks;
- Optical communications for data centers.

Finally, the last block is related to some emerging technologies, which are in intense research and possibly will be certain in the next future. Here, the features are as follows:

- Optical integration and silicon photonics;
- Optical wireless communication;
- Quantum communication.

3. The Zettabyte Era

In conclusion, from the first generation of optical systems to the most current, communication based on optical fibers has revolutionized the way people and the world are related. This evolution is constant and will continue to respond to people's increasing demand for data rates. The great development of the Internet era is also due to the evolution of the optical networks. The annual IP traffic is now measured in Exabyte. In order to understand this

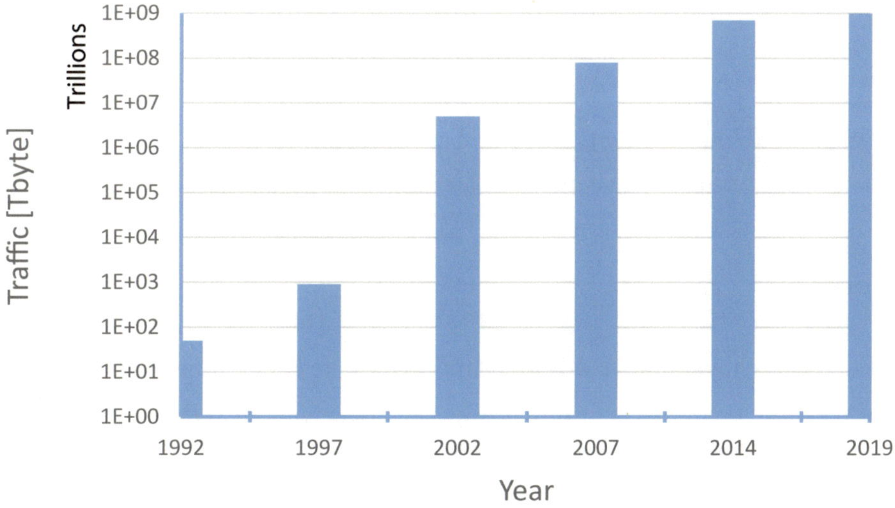

Figure 1. IP traffic: past, present, and future [2].

growth, **Figure 1** shows the past, present, and the predicted global Internet traffic according to Cisco's statistics [1, 2].

The bandwidth requirements are the cause for the rapid growth of Internet traffic over the past decade, and that growth has been exponential. We can expect that the bandwidth requirements in the second decade of this century will be ranging from 100 Mbps to 1 Gbps for residential access users, 10–40 Gbps for the majority of business users, and 100 Gbps to 1 Tbps for some institutions, such as government agencies or major research labs.

4. Conclusion

This book aims to introduce and treat one of the emerging technologies in the field of optical communications: optical wireless communication. This technology is a dynamic area of research and development because a growing demand by the end users for bandwidth in mobile communications to support broadband wireless services such as high definition TV mobile video, video conference high speed Internet access, and so on. As the global demand for bandwidth continues to increase, and the conventional radio frequency (RF) cellular microwave technologies have limitations in terms of bandwidth and spectrum, this solution based on optical wireless communication is one of the most promising alternative technologies for indoor and outdoor scenarios.

Another subject of interest in this book is related to optical hardware (lasers, modulators, optical amplifiers, and optical receivers), in order to improve the performance of microwave photonic links.

Author details

Pedro Pinho

Address all correspondence to: ppinho@deetc.isel.pt

ISEL—Instituto Superior de Engenharia de Lisboa, Lisbon, Portugal

References

[1] Cisco.The Zettabyte Era: Trends and Analysis. White Paper, June 2017

[2] Agrell E, et al. Roadmap of optical communications. Journal of Optic. 2016;**18**:063002 (40 pp). DOI: 10.1088/2040-8978/18/6/063002

Chapter 2

Challenges and Opportunities of Optical Wireless Communication Technologies

Isiaka Alimi, Ali Shahpari, Artur Sousa,
Ricardo Ferreira, Paulo Monteiro and
António Teixeira

Additional information is available at the end of the chapter

http://dx.doi.org/10.5772/intechopen.69113

Abstract

In this chapter, we present various opportunities of using optical wireless communication (OWC) technologies in each sector of optical communication networks. Moreover, challenges of optical wireless network implementations are investigated. We characterized the optical wireless communication channel through the channel measurements and present different models for the OWC link performance evaluations. In addition, we present some technologies for the OWC performance enhancement in order to address the last-mile transmission bottleneck of the system efficiently. The technologies can be of great help in alleviating the stringent requirement by the cloud radio access network (C-RAN) backhaul/fronthaul as well as in the evolution toward an efficient backhaul/fronthaul for the 5G network. Furthermore, we present a proof-of-concept experiment in order to demonstrate and evaluate high capacity/flexible coherent PON and OWC links for different network configurations in the terrestrial links. To achieve this, we employ advanced modulation format and digital signal processing (DSP) techniques in the offline and real-time mode of the operation. The proposed configuration has the capability to support different applications, services, and multiple operators over a shared optical fiber infrastructure.

Keywords: atmospheric turbulence, bit error rate (BER) performance, channel characterization, ergodic capacity, free-space optical (FSO) communication, relay-assisted transmission, RF/FSO technology, scintillation index

1. Introduction

The Internet is experiencing high growth with varieties of bandwidth-intensive mobile applications on an unprecedented scale. One of the potential reasons for the growth is the Internet of Things (IoT) technologies that have brought exceptional revolutions into the number of devices in the network. Conceptually, IoT entails ubiquitous existence of a variety of *things* such as mobile phones, sensors, actuators, and radio-frequency identification (RFID) tags. These entities are capable of interacting with each other as well as cooperating with their neighbors in order to accomplish common goals via unique addressing scheme [1]. It is envisaged that by the year 2020, billions of devices with an average of six to seven devices per person will be connected to the Internet [2]. The fifth generation (5G) wireless communication systems in which millimeter-wave (mm-wave) and massive multiple-input multiple-output (M-MIMO) antenna technologies are expected to be integrated are the promising solutions for supporting the huge amount of anticipated devices. However, the radio-frequency (RF)-based wireless mobile technologies transmission speeds are limited by the available RF spectrum in the regulated RF spectrum. This is due to various innovative wireless technologies and standards like WiMAX (IEEE 802.16), UWB (IEEE 802.15), Wi-Fi (IEEE 802.11), iBurst (IEEE 802.20), the cellular-based 3G and 4G [3]. Moreover, because of various advanced technologies being employed in the optical communications, there have been considerable advancements in the optical system capacity, network reach, and number of supported users. For example, the optical-fiber-based broadband network architectures like fiber to the home (FTTH) and fiber to the building (FTTB) present commercial solutions to the communication barriers by progressively rendering services closer to the customers via the passive optical network (PON) technologies such as gigabit PON (GPON), 10Gbps PON (XG-PON), and Ethernet PON (EPON). Currently, one of the major challenges is the capability to support various service requirements so as to achieve elastic and ubiquitous connections [4]. Consequently, convergence of wireless and optical networks is highly essential for cost-effective and pervasive network penetration for the next-generation network (NGN). The convergence will help in exploiting the mobility benefit offered by the wireless connectivity and the inherent bandwidth provided by the optical systems. This will help in achieving the anticipated capacity and energy-efficiency objectives of the NGNs [3]. Furthermore, optical wireless communication (OWC) system is one of attractive broadband access technologies that offer high speed as well as improved capacity. Consequently, the OWC can attend to the bandwidth requirements of different services and applications of the NGNs at relatively low cost [3, 5].

The OWC can be an alternative and/or complementary technology for the current wireless RF solutions. For instance, OWC operating at 350–1550 nm wavelength band can offer high data rate of about 30 Gb/s data rate. This advantage makes it an attractive solution for addressing the prevailing “last mile” and “last-leg” problems in the access network. Furthermore, in mobile communication, resources re-use is an important requirement in order to enhance the network coverage and capacity. OWC technology is able to meet this requirement with the aid of spatial diversity [5, 6]. OWC link can be of different configurations such as

1. Directed line of sight (LOS)
2. Nondirected LOS
3. Diffuse
4. Quasidiffuse
5. Multispot LOS

Out of these configurations, the LOS links have the highest data rates, lowest bit error rate (BER) performance, and less complex protocol. These features make LOS link the extensively employed configuration in the outdoor applications. Nevertheless, the major deficiencies of the LOS link are lack of mobility and susceptibility to blockage. The diffuse and nondirected LOS configurations, on the other hand, give better mobility advantages and are less susceptible to shading. However, noise, path loss, and multipath-induced dispersion relatively hinder their achievable data rate for high-speed links. Intensity modulation/direct detection (IM/DD) is the most widely used scheme in OWC systems. Furthermore, coherent scheme can also be employed to enhance channel usage. Implementation of coherent scheme relatively improves system performance at the expense of increased system complexity. This can be attributed to the fact that, precise wave-front matching between the incoming signal and the local oscillator (LO) is required to guarantee efficient coherent reception. Furthermore, DD application is uncomplicated as just low-cost transceiver devices are required without the necessity for the intricate high-frequency circuit designs relative to coherent systems [5, 6].

1.1. Block diagram of OWC system

A terrestrial OWC system consists of the transmitter, channel, and receiver. **Figure 1** illustrates a schematic of a terrestrial OWC system. The source at the transmitter generates information waveforms which are modulated onto an optical carrier. The optical field produced is then radiated over the atmospheric channel to the destination. The optically collected field at the receiver is then transformed to an electrical current. The detected electrical current is processed in order to recover the original transmitted information [8]. However, the received information may not be an exact replica of the original transmitted information because of the transmission loss experienced over the channel by the signal. This factor significantly limits the performance of wireless communications systems.

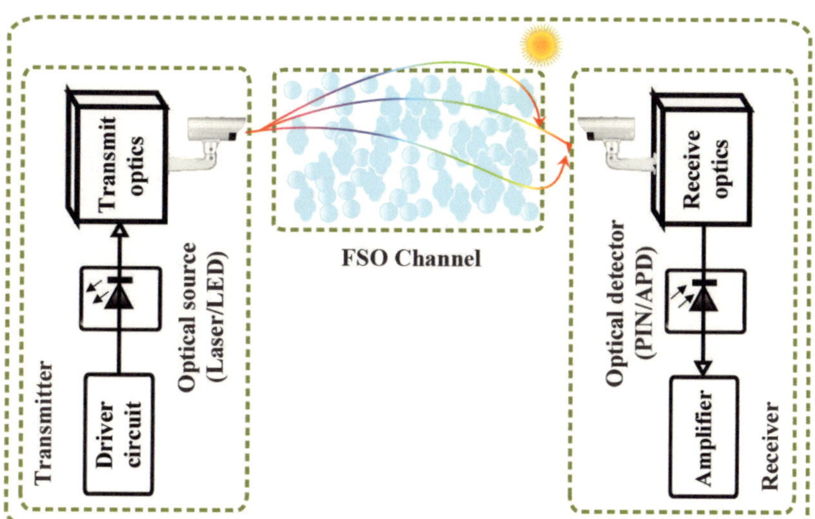

Figure 1. Block diagram of a terrestrial OWC system.

The transmission loss is mainly due to the resultant effects of scattering and absorption which are being introduced by the molecular constituents and aerosols along the transmission path. Therefore, scattering and wavelength-dependent absorption are the key components of atmospheric attenuation. Since absorption is a function of wavelength and wavelength selective, there are a range of wavelength windows that experience comparatively minimal absorptions. These transmittance windows in the absorption spectra of the atmospheric molecules are as shown in **Figure 2**.

In general, the wavelength ranges of 780–850 nm and 1520–1600 nm commonly used in the current OWC equipment are located in the atmospheric transmission windows where molecular absorption is negligible. This helps in mitigating the atmospheric absorption losses. Furthermore, certain wavelength windows that are located in the region of four specific wavelengths such as 850, 1060, 1250, and 1550 nm normally experience an attenuation of less than 0.2 dB/km. It is worth noting that the 850- and 1550-nm transmission windows coincide with the standard transmission windows of fiber communication systems. For this reason, majority of commercial OWC systems operate at these two windows in order to encourage the use of the available off-the-shelf components. Also, wavelengths like 10 μm and ultraviolet (UV) have also been considered for OWC systems. The 10-μm wavelength has better fog transmission characteristics, whereas the UV wavelength is more robust against impairment such as pointing errors and beam blockage. Also, the UV wavelength is less susceptible to solar and other background interferences [8].

Furthermore, it is worth noting that 1520–1600-nm wavelengths are compatible with erbium-doped fiber amplifier (EDFA) technology. This is highly essential in order to achieve high-power and high-data rate systems. Moreover, 1520–1600-nm wavelengths enable transmission of about 50–65 times more average output power than can be transmitted at 780–850 nm for a

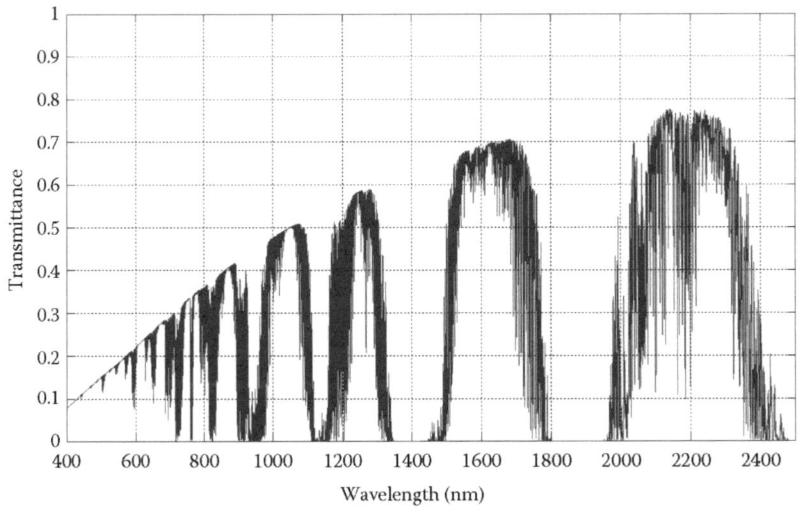

Figure 2. Atmospheric transmittance window with absorption contribution (adapted from Ref. [7]).

specified eye safety classification. This can be attributed to the low transmission of the human eye at these wavelengths [9].

1.2. Safety and regulations

One of the key factors for laser transmitter design is the safety issue. The infrared (IR) light sources can be likely safety threats to human if they are operated inappropriately. Also, exposure to certain optical beams may injure human skin and eye. However, the likely harm to the eye is comparatively more severe due to the eyes' ability to focus and concentrate optical energy. For instance, the eye can focus wavelength range of 0.4–1.4 μm on the retina with enough intensity to damage it; however, other wavelengths can to be absorbed by the front part of the eye before being focused on the retina. It should be noted that laser that is deemed to be "eye-safe" is also "skin-safe" [5, 9].

Moreover, it has been shown that the absorption coefficient at the front part of the eye is considerably greater for longer wavelengths (>1400 nm). Consequently, the permissible average transmission power for lasers operating at 1550 nm is relatively higher. Therefore, they are usually employed for longer transmission range [5, 9].

Figure 3 depicts the absorption of the eye at different wavelengths. At 700–1000-nm spectral range, the cost of optical sources and detectors are relatively low; however, the eye safety regulations are mainly strict. The maximum permissible exposure (MPE) at 900 nm wavelength is ~143 mW/sr. On the other hand, at longer wavelengths (≥1500 nm), the eye safety regulations are relatively less stringent; however, devices operating in these wavelengths are comparatively expensive. The guidelines on the safety of optical beams have been specified by several international standard organizations such as [5, 9]

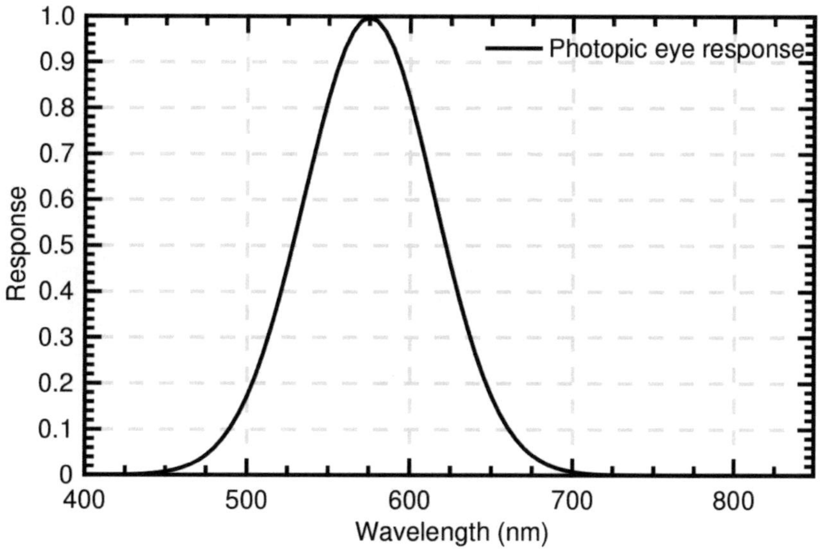

Figure 3. Response of the human eye at different wavelengths.

1. International Electrotechnical Commission (IEC)
2. Center for Devices and Radiological Health (CDRH)
3. European Committee for Electrotechnical Standardization (CENELEC)
4. American National Standards Institute (ANSI)
5. Laser Institute of America (LIA)

The aforementioned organizations have established mechanisms for categorizing lasers in accordance with their type and power. Generally, the classification is based on four groups which are Class 1 through Class 4. Comparatively, Class 1 is the least powerful whereas Class 4 is the most powerful. Also, each of the classes is specified by the accessible emission limits (AELs) metric. The AEL is determined by the optical source wavelength, the emitter geometry, and the source intensity [5, 9]. Consequently, the AEL varies from one OWC category to another. In the subsequent section, we present the major OWC categories.

2. OWC system classification

There have been growing research interests in the OWC system as a viable solution to attend to the NGN requirements in cost-effective ways. The two generic groups of OWC are indoor and outdoor optical wireless communications. The unlimited bandwidth offered by the OWC can be attributed to different bands such as IR, visible (VL), and UV being employed for communication purposes. **Figure 4** shows the electromagnetic spectrum for different applications. Furthermore, the spectrum illustrates the frequency and wavelength ranges being occupied by the bands in OWC. The indoor OWC employs IR or VL light for an in-building wireless solution. It is of high importance especially in scenarios in which the probability of offering network connectivity through physical wired connections is challenging. Moreover, the indoor OWC systems can be categorized into four broad configurations such as tracked, diffused, nondirected LOS, and directed line of sight (LOS). Furthermore, the outdoor OWC employs optical carrier for transporting information from one point to another over an unguided channel that could be an atmosphere or a free space. So, this OWC technology is also known as a free-space optical (FSO) communication system. The FSO communication systems operate

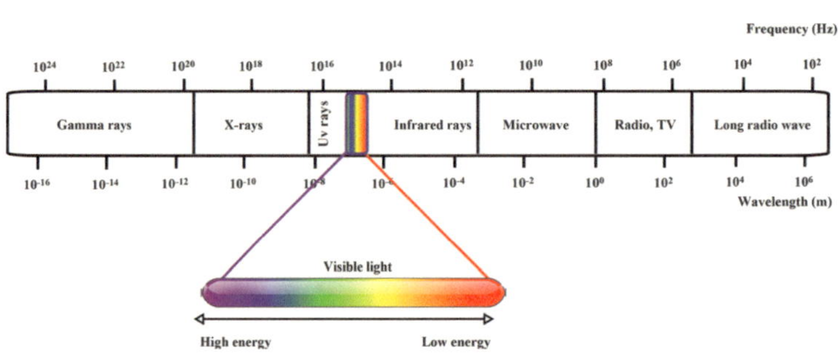

Figure 4. Electromagnetic spectrum.

at the near IR frequencies and are classified into terrestrial and space optical links. These consist of building-to-building, satellite-to-ground, ground-to-satellite, satellite-to-satellite, and satellite-to-airborne platforms (unmanned aerial vehicles [UAVs] or balloons) [10]. The tree diagram in **Figure 5** shows the OWC system classification.

2.1. Underwater optical wireless communications (UOWCs)

The underwater wireless communications are a process of transmitting data in unguided water environments via wireless carriers such as acoustic wave, RF wave, and optical wave. Compared to the RF or acoustic alternatives, UOWC offers higher data rate and transmission bandwidth. Basically, the UOWC uses optical wave as wireless carrier for an unguided data transmission. The UOWC systems are applicable in disaster precaution, offshore exploration, environmental monitoring, as well as military operations. Nevertheless, UOWC systems are susceptible to absorption and scattering which are normally created by the underwater channels. These conditions lead to severe attenuation of optical wave and eventually hindered the system performance. Different viable techniques have been presented in the literature to attend to the associated technical challenges of a UOWC. One of such is an underwater wireless sensor network (UWSN). **Figure 6** depicts a UWSN with aerospace and terrestrial communications.

The major entities in the UWSN are distributed nodes such as relay buoys, seabed sensors, autonomous underwater vehicles (AUVs), and remotely operated underwater vehicles (ROVs). The network entities are capable of performing tasks like processing, sensing, and communication in order to sustain collaborative monitoring of the underwater environment. The acquired data by the sensors that are located at the seabed are transmitted through acoustic/optical links to the AUVs and ROVs which in turn relay the signal to the ships, communication buoys, and other underwater vehicles. Furthermore, the onshore data center that is above the sea surface then processes the data and communicates with the satellite and the ships via RF/FSO links [11].

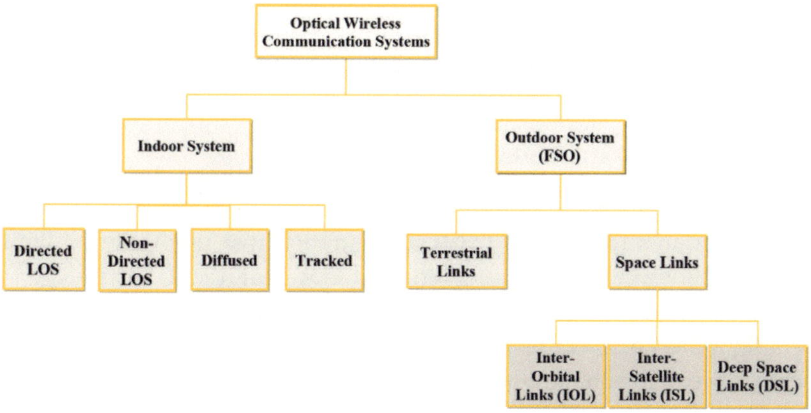

Figure 5. Optical wireless communication system classification (adapted from Ref. [10]).

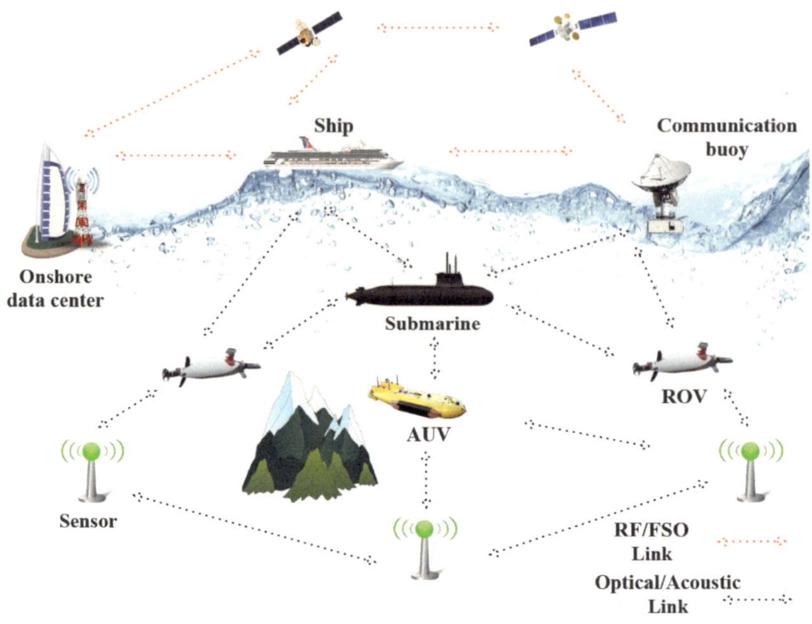

Figure 6. UWSN with aerospace and terrestrial communication.

2.1.1. UOWC link configuration

The four categories of UOWC based on the link configurations of the UWSN's nodes are as follows:

1. **Point-to-point line-of-sight (PTP-LOS) configuration**

   The PTP-LOS arrangement is the widely adopted link configuration in the UOWC. In this configuration, the receiver detects the light beam in the direct path to the transmitter. Due to the fact that light sources like lasers are generally used in the PTP-LOS-based systems, precise pointing between the transceiver is essential. Hence, the requirement limits the performance of UOWC systems in turbulent water environments. Also, this can even be more stringent when the transmitter and the receiver of the underwater vehicles are not stationary [11].

2. **Diffused LOS (D-LOS) configuration**

   Unlike the PTP-LOS scheme that employs light sources with narrow divergence angle, the D-LOS is a point-to-multipoint (PTMP) configuration that uses diffused light. The light sources like high-power light-emitting diodes (LEDs) that have large divergence angle are used for the UOWC transmission from one node to multiple nodes. The broadcast nature of this configuration helps in relaxing the requirement for precise pointing. Nonetheless, the scheme is prone to aquatic attenuation which is as a result of the large interaction area with the water. Consequently, this limits D-LOS application to comparatively short distances and lower data rate communications [11].

3. **Retroreflector-based LOS (R-LOS) configuration**

   The R-LOS configuration is a special form of PTP-LOS scheme. The R-LOS scheme is appropriate for duplex UOWC systems in which the sensor nodes are expected to have low form factors and consume less power. Due to the fact that there is no laser or other light sources in the retroreflector end, its power consumption, weight, and volume are significantly reduced. However, the R-LOS performance is limited by the backscatter of the transmitted optical signal that may interfere with the reflected signal [11].

4. **Non-line-of-sight (NLOS) configuration**

   Another category of UOWC that does not require alignment restriction of LOS is NLOS configuration. The concept of this configuration is that the transmitter launches the beam of light to the sea surface in such a way that the angle of incidence is greater than the critical angle. This ensures that the light beam experiences a total internal reflection. In this scheme, the receiver is directed toward the sea surface in a direction approximately parallel to the reflected light for appropriate signal reception. However, signal dispersion may occur due to the random sea surface slopes (i.e., induced by wind or other turbulence sources) that may reflect the light back to the transmitter [11].

2.1.2. Advantages and challenges of UOWC

The three means of realizing an underwater wireless communication for UWSNs are acoustics, RF, and optics [11]. We compare all schemes so as to point out some salient features and advantages of UOWC:

1. **Acoustic communication**

   Acoustic communication is the widely employed method in the underwater wireless communication due to the fact that it is the most viable means of realizing the longest underwater link range. However, the frequency range of operation of acoustic communication is limited (between tens of hertz and hundreds of kilohertz). So, comparatively, this results in low transmission rate in the order of Kbps. Furthermore, the speed of sound wave in the fresh water is about 481 m/s at 2°C and in the salt water it is about 1500 m/s (1500.235 m/s at 1 MPa, 10°C and 3% salinity). These comparative slow propagation speeds make acoustic links prone to serious communication delay. Consequently, acoustic communication cannot be employed in applications that require large volume of data transmission in real time. Additionally, acoustic technology not only impacts marine life but also involves the use of costly, bulky, and energy-consuming transceivers [11].

2. **RF communication**

   There are two main advantages of underwater RF communication. In relation to other UWC methods, the RF wave can achieve a smooth transition through the air/water interface. The feature enables a cross-boundary communication that unifies both terrestrial and underwater RF communication systems. It is remarkable that underwater RF communication is the method that is most tolerant to the water turbulence and turbidity. Nevertheless, the method is limited by the associated short link range. Likewise, the underwater RF methods are cost-ineffective due to the required high energy-consuming transceivers, costly and huge transmission antenna [11].

3. **Optical communication**

The outstanding technical merits of UOWC are the lowest link delay, highest communication security, highest transmission rate, and lowest implementation costs compared to other methods. The comparative high-speed benefit of UOWC makes it a promising candidate for real-time applications like underwater video transmission. In UOWC systems, LOS configuration is normally employed in the transmission. This helps in preventing eavesdroppers, and hence improves communication security. Besides, UOWC is the most cost-effective and energy efficient means of underwater wireless communication. This can be attributed to the comparatively small, low-energy consuming, and low-cost optical transceivers which are normally employed. This aids in system scalability and large-scale commercialization of UOWC [11]. However, UOWC also has certain associated challenges, which makes its realization highly demanding. The key challenges of UOWC are enumerated as follows:

a. **Poor system performance**

When light propagates in water, the photon interacts with the water molecules as well as other particulate matters inside the water. The interaction results in absorption and scattering which attenuate the transmitted optical signal and eventually bring about multipath fading. Consequently, in turbid water environment, UOWC experiences poor BER performance over just a few hundred meters link distance. Furthermore, the presence of matters like chlorophyll and other colored dissolved organic material (CDOM) leads to an increase in the turbidity of the water. So, they reduce the propagation distance of the transmitted optical signal. Besides, since the concentration of CDOM varies with ocean depth variations, the associated light attenuation coefficients also change. This anomaly results in an increase in the system complexity [11, 12].

b. **Intermittent connection**

In practice, owing to the narrow divergence feature of blue/green lasers or LEDs, they are usually employed as the light sources in UOWC systems. Therefore, the narrow feature demands a precise alignment between the optical transceivers. However, there is high tendency for intermittent misalignment of optical transceivers because of the turbulence in the underwater environment. Hence, sporadic movements of the sea surface will bring about severe connectivity loss [11, 12].

c. **Energy constraint and system reliability**

Factors such as the pressure, temperature, flow, and salinity of the seawater have high impact on the performance and lifetime of UOWC devices. Furthermore, UOWC devices are susceptible to failures due to corrosion and fouling. So, reliable underwater devices are required for effective communications. Also, due to the fact that exploitation of solar energy is extremely difficult and UOWC devices are expected to experience extended operation time, then, reliability of the batteries and power consumption efficiency of device are essential in a UOWC environment [11, 12].

2.2. Visible light communication (VLC) systems

The current enhancement of LED chip design with swift nanosecond-switching times and extensive deployment of LEDs for energy efficiency paves the way for visible light communication (VLC) system [13, 14]. So, the VLC system has become an attractive technology for addressing challenges such as energy efficiency, bandwidth limitation, electromagnetic radiation, and safety in wireless communications [15, 16]. The VLC system operates in the wavelength range of ~390–750 nm. **Table 1** shows the seven colors with specific frequency and wavelength for visible spectrum. **Figure 7** illustrates VLC system implementation. The concurrent support for communication and illumination by the VLC offers the following advantages over the RF communications.

Color	Wavelength (nm)	Frequency (THz)
Violet	440–405	680–740
Indigo	455–440	660–680
Blue	475–455	630–660
Green	560–475	535–630
Yellow	585–560	515–535
Orange	605–585	495–515
Red	685–605	435–495

Adapted from Ref. [17].

Table 1. Visible spectrum colors.

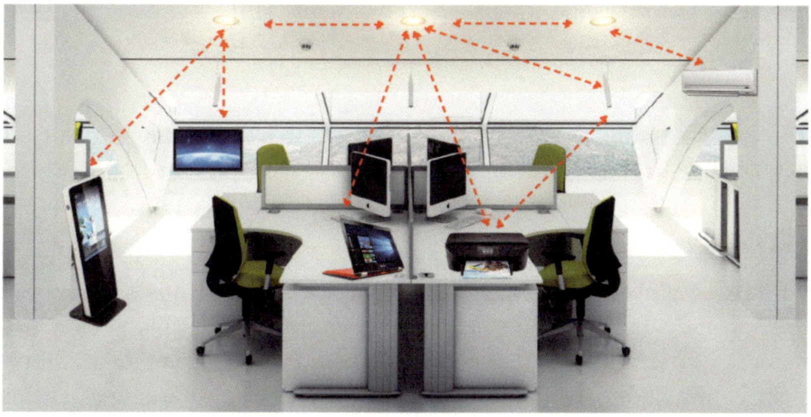

Figure 7. Visible light communication system.

2.2.1. *Huge bandwidth*

It exhibit almost unlimited and unlicensed bandwidth which approximately ranges from 380 to 780 nm. Therefore, VLC has 350 THz that can support multi-gigabit-per-second data rates with LED arrays in a multiple-input multiple-output (MIMO) configuration [14]. This makes VLC a good alternative to the indoor IR that operates at 780–950 nm for the access technologies [6].

2.2.2. *Low power consumption*

VLC provides both communication and lighting, giving Gbps data rates with only unsophisticated LEDs and photodetectors (PDs) that consume low power compared to costly RF alternatives that demand high power consumption for sampling, processing, and transmitting Gbps data [14].

2.2.3. *Low cost*

The required optical components such as LEDs and photodetectors are inexpensive, compact, lightweight, amenable to dense integration, and have very long lifespan [13]. Moreover, with large unlicensed optical spectrum as well as much lower power-per-bit cost compared to the RF communications, VLC is relatively cheaper.

2.2.4. *No health concerns*

VLC does not generate radiation that leads to public health concern. Besides, it lowers the carbon dioxide emission owing to the little extra power consumption for communication purposes [18].

2.2.5. *Ubiquitous computing*

Due to the fact that there are various luminous devices like traffic signs, commercial displays, indoor/outdoor lamps, TVs, car headlights/taillights, and so on being used everywhere, VLC can be employed for a wide range of network connectivity [19].

2.2.6. *Inherent security*

VLC offers comparatively higher security due to the fact that it is highly intricate for a network intruder that is outside to pick up the signal [14].

2.2.7. *Indoor localization*

The existing RF-based global positioning system (GPS) gives inadequate or no network coverage in the indoor and underground (e.g., tunnel) environment. This is as a result of high attenuation, multipath, and the safety regulation. These factors lead to an accuracy of only up to a few meters for the RF-based GPS. The VLC-based indoor positioning can be employed to attend to the issues in the enclosed environments. So, the VLC-based indoor navigation services offer very high accuracy to within a few centimeters. In essence, VLC offers good indoor localization system using the white LEDs. Furthermore, LEDs give better light source that is more than 400 lux. This is sufficient for high-speed data transmission compared to the incandescent and fluorescent sources. Moreover, LEDs have longer lifespan that results in ecological and financial benefits [6].

In addition, VLC is an alternative technology in sensitive or hazardous environments like airplanes, hospitals, and industrial gas production plants where the employment of RF technology is not permitted. The worldwide research community through bodies like the Wireless World Research Forum, the IEEE standardization body, the VLC consortium, the UK research council, and the European OMEGA project has embraced the indoor short-range VLC as a promising scheme because of the associated excellent attributes [6].

However, simultaneous employment of light sources for the data communication as well as illumination causes certain challenges that require consideration for VLC system implementations to be viable. Flicker mitigation and dimming support are the two major challenges of visible light spectrum [14, 20]. In the contemporary lighting systems, the light sources are equipped with dimming control functionality that enables the users to control the average brightness of light sources to their preferred level. Flicker is the variation in the brightness of light perceived by the human naked eye. Flicker is as a result of continuous switching on and off of the light source during data transmission. It is essential to mitigate any possible flicker because it can instigate negative/harmful physiological changes in humans. Flicker can be prevented by making the changes in brightness be within the maximum flickering time period (MFTP). The MFTP is the maximum time period within which the light intensity can be changed without any perception by the human eye [14]. Also, various modulation formats have been recommended for VLC in view of dimming control and flicker mitigation. For instance, the IEEE 802.15.7 standard proposes variable pulse position modulation (VPPM) for VLC system because of its notable ability to control dimming. It integrates the pulse position modulation (PPM) and pulse width modulation (PWM) in order to support communication with dimming control [14, 20]. Furthermore, other challenges are high path losses, multipath-induced intersymbol interference (ISI), artificial light-induced interference, and blocking. Moreover, LED electro-optic response nonlinearity has to be taken into consideration [6].

2.3. Wireless body area network (WBAN)

Wireless body area network (WBAN) is a system that comprises a set of miniaturized low-power, lightweight sensor nodes, which form wireless sensor networks (WSNs). **Figure 8** shows a WBAN system for medical monitoring. With the help of the sensor nodes, WBANs have emerged as an attractive alternative to the conventional wired medical network. Also, there has been noteworthy increase in the WBAN systems because of the IEEE 802.15.6 standard that regulates their commercial applications. Furthermore, there is an obligation on 2.36–2.40-GHz frequency band as a medical-only WBAN band by the Federal Communications Commission (FCC). The restriction is purposely for service provisions for the indoor health-care facilities as well as for supporting the patients' health-care information and management [21, 22]. Furthermore, the constraint is also to guarantee high quality of service for the health information transmission.

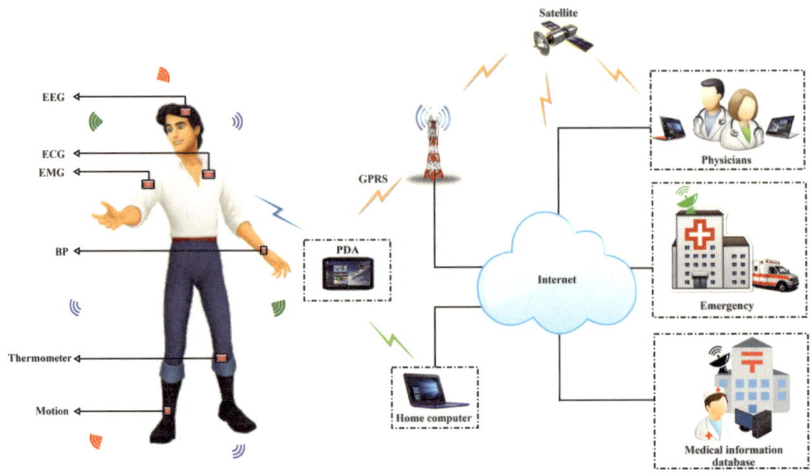

Figure 8. Wireless body area network.

The distributed sensors in the WBAN can be implanted in or on the human body in order to monitor physiological parameters in real time. The implanted sensors that are wirelessly connected to the outside network through a central unit collect different vital health information. The monitored physiological data include electromyogram (EMG), electroencephalogram (EEG), electrocardiogram (ECG), temperature, heart rate, blood pressure (BP), and glucose level [22]. The ZigBee, Bluetooth, as well as the current Bluetooth Low Energy (BLE) are contending for market share of wireless health devices. Their major appealing advantages are low power consumption and the added mobility [23]. Health devices that employ these technologies are operating in the industrial, scientific, and medical (ISM) radio bands. High emissions from these devices can create electromagnetic interference (EMI) and eventually disrupt communication. Similarly, there are security issues concerning data transmission for patient monitoring which are susceptible to hacking [23].

Furthermore, it should be noted that the existing WBANs that use ultra-wide band (UWB) transmissions are RF based. However, their implementation in the hospitals and medical facilities where RF-based system deployment is restricted or prohibited can be challenging. This is due to the potential effects of EMI from various RF transceivers on medical devices. The EMI effects can lead to medical equipment malfunction. Also, RF wave propagation on and/or in the human body is highly complex to examine. Consequently, to address these challenges OWC can be employed as an alternative solution [24].

The ECG signal and patient information can be transmitted concurrently with the help of VLC technology. Moreover, certain medical equipment like the one for the cardiac stress test (or cardiac diagnostic test) can be improved on by incorporating LEDs on the sensor units. This implementation will help in minimizing the large amount of cables (e.g., electrodes) that are normally required. Besides, VLC employment is greener (green communication and networking), safer, more secure in RF-restricted/prohibited hospitals and medical facilities [23]. Furthermore, it is worth noting that the current advancements in the organic LED (OLED) technology enable the integration of VLC transceivers into wearable gadgets and clothing [24].

2.4. Optical space communications

An effective communication links between satellites enable better flexibility, extended coverage, and improved connectivity to be achieved in satellite systems. This is applicable in interorbit links between satellites in low earth orbit (LEO) and geostationary orbit (GEO). Furthermore, intersatellite links between satellites in the same GEO or LEO orbit are other scenarios for application. Also, the satellite system connectivity can be enhanced by exploiting the free-space links between satellites. This will result in an improved capacity for telecommunication systems. Moreover, the capability to relay data from the earth observation satellite to the ground through a GEO relay satellite enables real-time data flow and minimizes the number of ground stations required for service delivery in the system [25].

Generally, the space link implementation can be realized at microwave, millimetric, and optical frequencies. Also, all technologies require the communication beam that emanates from the transmitting terminal to be pointed toward the receiving terminal with sufficient accuracy. This is necessary in order to meet the required link power budget. **Table 2** presents typical antenna gains and beamwidths for intersatellite link technologies [25].

It should be noted that out of the technologies, optical frequency offers an exceptionally high antenna gain for comparatively small antenna size. Therefore, optical links provide significant advantages such as low mass, low power consumption, and reduced size. As demonstrated in [26], the optical antenna diameter may be relatively smaller than that of RF by a factor of 13. Also, its mass and power can be half of that for RF systems. **Table 3** compares different communication technologies in terms of mass, power, and antenna diameter. An architecture that depicts ground-to-satellite optical links that are connected to satellite network and then satellite-to-ground optical links is illustrated in **Figure 9**. The salient features of optical links enable small terminals that can be easily accommodated on the satellite [25]. Also, optical space-based communication offers high-data rate, large capacity, minimized interference risk with other communication systems, and efficient utilization of frequency resources [10, 26].

Nevertheless, the space FSO communications are susceptible to atmospheric effects which make the channel a random function of space and time. Consequently, the uplink/downlink transmissions suffer from losses such as atmospheric scintillation, beam divergence, absorption, scattering, misalignment, cloud blockage, background noise and angle-of-arrival fluctuations. It should be noted that the losses confronted by the beam in the FSO uplink transmission

Band	Parameters			
	Frequency	Antenna diameter (m)	Antenna gain (dBi)	3 dB beamwidth degrees
S-band	2 GHz	2	30	5.25
Ka-band	26 GHz	2	52	0.40
Millimeter	60 GHz	1	53	0.35
Optical	0.36 PHz	0.07	108	6.7×10^{-4}

Adapted from Ref. [25].

Table 2. Intersatellite link technologies comparison.

Link	Frequency band								
	Optical			*Ka*			Millimeter		
	Antenna diameter	Mass	Power	Antenna diameter	Mass	Power	Antenna diameter	Mass	Power
GEO-LEO	10.2 cm (1.0)	65.3 kg (1.0)	93.8 W (1.0)	2.2 m (21.6)	152.8 kg (2.3)	213.9 W (2.3)	1.9 m (18.6)	131.9 kg (2.0)	184.7 W (2.0)
GEO-GEO	13.5 cm (1.0)	86.4 kg (1.0)	124.2 W (1.0)	2.1 m (15.6)	145.8 kg (1.7)	204.2 W (1.6)	1.8 m (13.3)	125.0 kg (1.4)	175.0 W (1.4)
LEO-LEO	3.6 cm (1.0)	23.0 kg (1.0)	33.1 W (1.0)	0.8 m (22.2)	55.6 kg (2.4)	77.8 W (2.3)	0.7 m (19.4)	48.6 kg (2.1)	68.1 W (2.1)

Adapted from Ref. [26].

Table 3. Comparison of different technologies with transmit power of 10, 50, and 20 W for optical-, Ka-, and millimeter-band systems, respectively.

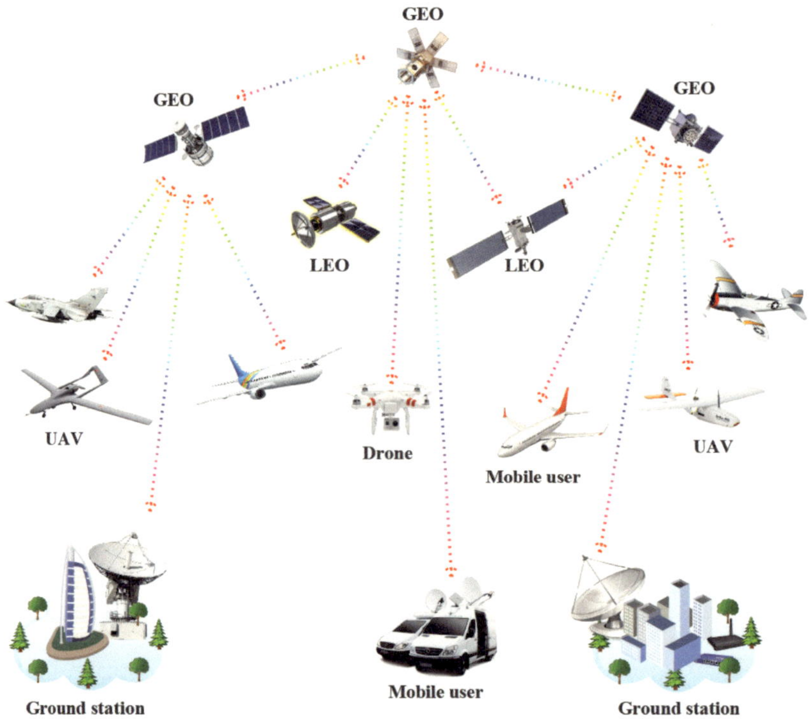

Figure 9. Space optical communication.

are very large compared to downlink transmission. Furthermore, the intersatellite FSO links are insusceptible to weather conditions due to the fact that the satellite orbits are at a considerable distance over the atmosphere [10]. However, the associated narrow optical beamwidths bring about stringent pointing, acquisition, and tracking (PAT) requirements for optical systems. This can also be attributed to different relative velocity of the satellite. So, a compact low-mass PAT system is required to address PAT requirements [26].

2.5. Terrestrial free-space optical (FSO) communications

There have been much more research effort in terrestrial FSO partly because of some successful field trials and commercial deployments [27–31]. **Figure 10** shows a scenario for FSO system deployment as a universal platform for a nippy as well as an efficient ubiquitous wireless service provision for the future broadband access networks. The significant attentions being attracted by the FSO systems are primarily due to their inherent advantages such as cost-effectiveness, lower power consumption (high-energy-efficiency-green communication), ease of deployment, higher bandwidth/capacity, more compact/low-mass equipment, reduced time-to-market, immunity to EMI, high degree of security against eavesdropping, license-free operation, as well as better protection against interferences, compared with the traditional RF communication systems [32, 33]. These salient features make FSO communication systems very appealing for a variety of applications in disaster recovery, radio astronomy, remote-

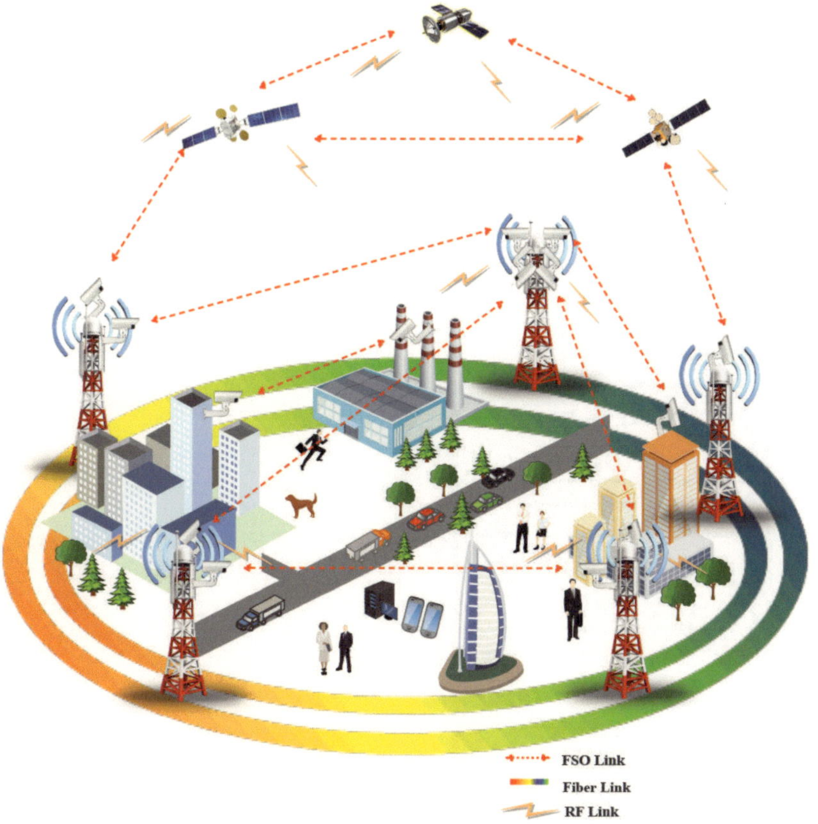

Figure 10. A scenario for OWC system deployment for access networks.

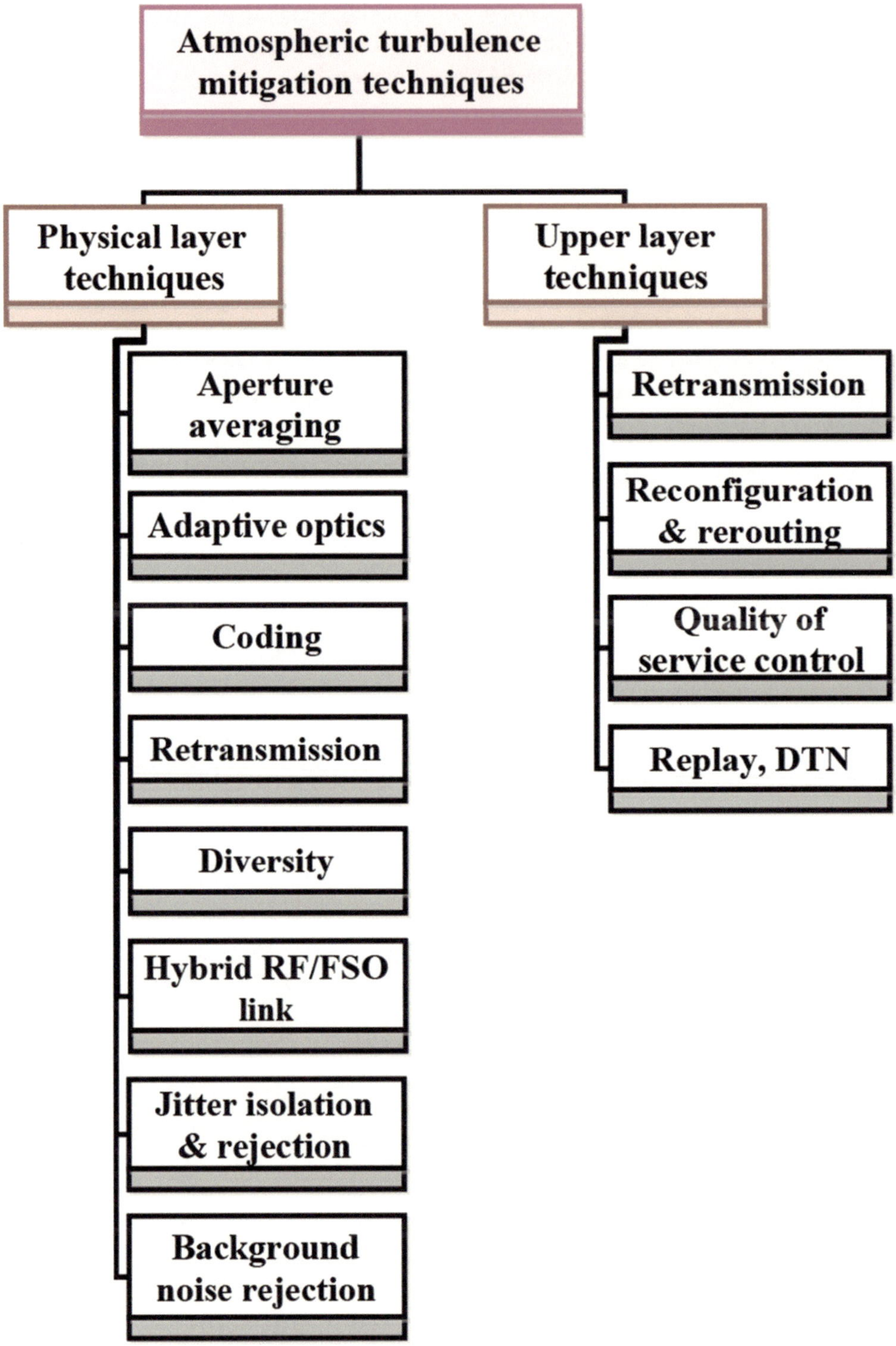

Figure 11. Atmospheric turbulence mitigation techniques (adapted from Ref. [10]).

sensing/surveillance/monitoring, metropolitan area network extension, high-definition TV transmission, sharing of medical imaging in real time, and fronthaul and backhaul for wireless cellular networks [3, 8]. Moreover, apart from being used for establishing terrestrial links, they are applicable for launching high-speed interplanetary space links such as intersatellite/deep space and ground-to-satellite/satellite-to-ground links [10].

In spite of the advantages of FSO communication and diverse application, its extensive use is hindered by some challenges in real-life scenarios. For instance, the FSO links are susceptible to scattering caused by adverse weather conditions like snow, rain, and fog [8, 33]. Moreover, building sway caused by factors such as thermal expansion, wind loads, and weak earthquakes also impairs the FSO link performance [33]. Also, atmospheric turbulence-induced fading has been recognized as the main contributor of the FSO link impairment [33, 34]. Consequently, the system performance is impeded as a result of atmospheric effects which cause loss of spatial coherence, beam spreading, and temporal irradiance fluctuation known as scintillation or fading [8, 35]. Scintillation manifests as temporal and spatial variation in the light intensity along the transmission path. This is due to the random changes in the refractive index which is as a result of the inhomogeneities in the air temperature and pressure [8, 32]. For these reasons, the dispersive nature of the link needs considerable attention when modeling an FSO system. This will help in supporting the stringent requirements of various bandwidth-intensive mobile applications of the NGNs [3]. The FSO link impairment modelings are discussed in Section 3. The tree diagram in **Figure 11** presents different atmospheric turbulence mitigation techniques.

3. Optical system and channel model

Assuming a practical FSO link with intensity-modulation/direct-detection (IM/DD) using OOK modulation, the data are modulated onto the instantaneous intensity of an optical beam at the transmitter. The optical power emanated from the transmit aperture into the free space is assumed to be affected by factors such as misalignment fading (pointing errors), atmospheric turbulence-induced fading, and background noise or ambient noise before reaching the receive aperture. These factors lead to the signal intensity fluctuation. Consequently, the subsequent received electrical signal, r, at the receive aperture can be modeled as [3, 35–37]

$$r = \eta_e hx + n, \tag{1}$$

where $x \in \{0,1\}$ denotes the transmitted information bit, η_e is the effective photoelectric conversion ratio of the receiver, n is additive white Gaussian noise (AWGN) with zero mean and variance, $\sigma_n^2 = N_0/2$,N_0 is a one-sided noise power spectral density in watts/Hz, and $h = h_\ell h_a h_p$ represents the irradiance that influences the channel state. The channel irradiance in Eq. (1) is a product of the deterministic path loss, h_ℓ: the random attenuation (i.e., atmospheric turbulence-induced fading), h_a, as well as the random attenuation (due to geometric spread and pointing errors), h_p. The h_a and h_p are random variables with probability density functions (pdfs) $f_{h_a}(h_a)$ and $f_{h_p}(h_p)$, respectively.

3.1. Atmospheric attenuation

When an optical beam passes through the atmosphere, it experiences atmospheric loss. The attenuation suffered by the signal power according to the exponential Beers-Lambert law is given by [36–38]

$$h_\ell(\lambda, z) = \frac{P(\lambda, z)}{P(\lambda, 0)} = \exp\left(-\sigma(\lambda)z\right) \tag{2}$$

where $h_\ell(\lambda, z)$ denotes the loss that is a function of propagation path of length z at wavelength λ, $P(\lambda, z)$ and $P(\lambda, 0)$ are the signal power and the emitted power at distance z, respectively. The attenuation coefficient or the total extinction coefficient, $\sigma(\lambda)$ per unit of length, is given by [36]

$$\sigma(\lambda) = \alpha_m(\lambda) + \alpha_a(\lambda) + \beta_m(\lambda) + \beta_a(\lambda), \tag{3}$$

where $\alpha_{m,a}$ represent molecular and aerosol absorption coefficients, respectively, and $\beta_{m,a}$ denote molecular and aerosol scattering coefficients, respectively.

The attenuation h_ℓ is assumed to be a constant scaling factor over a long time period, so, there is no haphazard behavior. Moreover, it is subject to the distribution and size of the scattering particles as well as the wavelength employed. It can be expressed in terms of visibility, which can be measured directly from the atmosphere. Empirically, attenuation can be defined in terms of visibility as [36]

$$\sigma(\lambda) = \frac{3.912}{V}\left(\frac{\lambda}{550}\right)^{-q}, \tag{4}$$

where V is the visibility (in kilometers) and q is a parameter that depends on the particle size distribution in the atmosphere expressed by the Kruse model as [36]

$$q = \begin{cases} 1.6 & V > 50 \text{ km} \\ 1.3 & 6 \text{ km} < V < 50 \text{ km} \\ 0.585V^{1/3} & V < 6 \text{ km} \end{cases} \tag{5}$$

Furthermore, Kim presents an extended model in order to achieve a better accuracy at lower visibility scenarios. The Kim model is expressed as [36]

$$q = \begin{cases} 1.6 & V > 50 \text{ km} \\ 1.3 & 6 \text{ km} < V < 50 \text{ km} \\ 0.16V + 0.34 & 1 \text{ km} < V < 6 \text{ km} \\ V - 0.5 & 0.5 \text{ km} < V < 1 \text{ km} \\ 0 & V < 0.5 \text{ km} \end{cases} \tag{6}$$

3.2. Pointing error or misalignment fading

An FSO link is an LOS communication with narrow optical beamwidth that brings stringent pointing accuracy requirements for efficient performance and reliability of optical systems.

Pointing errors and signal fading normally occur at the receiver because of the wind loads and thermal expansions that lead to random building sways. Assuming a Gaussian spatial intensity profile of beam waist w_z at the receiver plane, located at distance z away from the transmitter. Also, suppose a circular aperture of radius r, the collected power fraction owing to the geometric spread with radial displacement α, from the detector origin can be approximated as the Gaussian form [37, 38]

$$h_{\mathrm{p}}(\alpha) \approx A_0 \exp\left(-\frac{2\alpha^2}{w_{z_{eq}}^2}\right) \tag{7}$$

where $w_{z_{eq}}^2 = w_z^2\sqrt{\pi}\mathrm{erf}(v)/2v\exp(-v^2)$, $v = \sqrt{\pi}r/\sqrt{2}w_z$, $A_0 = [\mathrm{erf}(v)]^2$, $w_{z_{eq}}$ is the equivalent beamwidth, and $erf(\cdot)$ denotes the error function given by [39]

$$\mathrm{erf}(x) = (2/\sqrt{\pi})\int_0^x e^{-u^2}du \tag{8}$$

If we assume that the elevation and the horizontal sway are independent and identical Gaussian distributions, then the radial displacement, α, follows a Rayleigh distribution. Then, the $f_{h_{\mathrm{p}}}(h_{\mathrm{p}})$ can be defined as [37, 38]

$$f_{h_{\mathrm{p}}}(h_{\mathrm{p}}) = \frac{\gamma^2}{A_0^{\gamma^2}} h_{\mathrm{p}}^{\gamma^2-1}, \quad 0 \le h_{\mathrm{p}} \le A_0 \tag{9}$$

where $\gamma = w_{z_{eq}}/2\sigma_s$ represents the ratio between the equivalent beam radius and the standard deviation (jitter) of the pointing error displacement at the receiver and σ_s^2 is the jitter variance at the receiver.

3.3. Atmospheric turbulence

The intensity fluctuation over the FSO channel has been defined by several statistical models in the literature for different turbulence regimes. One of such model is the log-normal (LN) distribution that has been extensively employed because of its significant match with the experimental measurements. This is discussed further in Section 6. Some of other widely employed models are gamma-gamma (ΓΓ), negative exponential, *K* distribution, and *I-K* distribution. However, in this work, we focus on the LN, ΓΓ, and generic Málaga ($\mathcal{M}$)-distribution models.

3.3.1. Log-normal distribution (LN)

In general, the LN model is only suitable for weak turbulence conditions and for link range that is less than 100 m [40]. So, the intensity fluctuation pdf for the weak turbulence modeled by the LN distribution is given by [3, 35]

$$f_{h_a}(h_a) = \frac{1}{2h_a\sqrt{2\pi\sigma_x^2}} \exp\left(-\frac{(\ln(h_a) + 2\sigma_x)^2}{8\sigma_x^2}\right), \tag{10}$$

where $\sigma_x^2 = \sigma_l^2/4$ represents the log-amplitude variance defined for plane wave and spherical waves, respectively as [3, 35]

$$\sigma_x^2|_{\text{plane}} = 0.307C_n^2 k^{7/6} L^{11/6}, \tag{11}$$

$$\sigma_x^2|_{\text{spherical}} = 0.124C_n^2 k^{7/6} L^{11/6}, \tag{12}$$

$$\sigma_l^2|_{\text{plane}} = 1.23C_n^2 k^{7/6} L^{11/6}, \tag{13}$$

$$\sigma_l^2|_{\text{spherical}} = 0.50C_n^2 k^{7/6} L^{11/6}, \tag{14}$$

where σ_l^2 is the log-irradiance variance, $k = 2\pi/\lambda$ represents the optical wave number, L denotes the distance, and C_n^2 is the altitude-dependent index of refraction structure parameter. The C_n^2 is a main parameter for distinguishing the amount of refractive index fluctuation in the atmospheric turbulence. It is a function of the atmospheric altitude, wavelength, and temperature. There are a number of C_n^2 profile models being proposed in the literature; however, the extensively used one in terms of altitude is the Hufnagle-Valley model given by [3, 5, 35, 39]

$$C_n^2(h) = 0.00594(v_w/27)^2(10^{-5}h)^{10}\exp(-h/1000) + 2.7 \times 10^{-16}\exp(-h/1500) + \hat{A}\exp(-h/100), \tag{15}$$

where h signifies the altitude in meters (m) and $\hat{A}$ represents the nominal value of $C_n^2(0)$ at the ground level in $m^{-2/3}$. The value of C_n^2 for the FSO links near the ground level is approximately $1.7 \times 10^{-14} m^{-2/3}$ and $8.4 \times 10^{-14} m^{-2/3}$ during the daytime and at the night, respectively. Generally, C_n^2 ranges from $10^{-13} m^{-2/3}$ for the strong turbulence to $10^{-17} m^{-2/3}$ for the weak turbulence. Its typical average value is $10^{-15} m^{-2/3}$ [5, 39]. The v_w denotes the root mean square (rms) wind speed (pseudowind) in meters per second (m/s) with the most usual value of 21 m/s, but it can be described by [39]

$$w = \left[\frac{1}{15 \times 10^3}\int_{5\times10^3}^{20\times10^3} V^2(h)dh\right]^{1/2}, \tag{16}$$

where *V*(*h*) is normally defined by the Bufton wind model that can be expressed as [39]

$$V(h) = \omega_s h + V_g + 30\exp\left[-\left(\frac{h - 9400}{4800}\right)^2\right] \tag{17}$$

where V_g denotes the ground wind speed and ω_s signifies the *slew rate* that is related to the satellite movement regarding the observer on the ground.

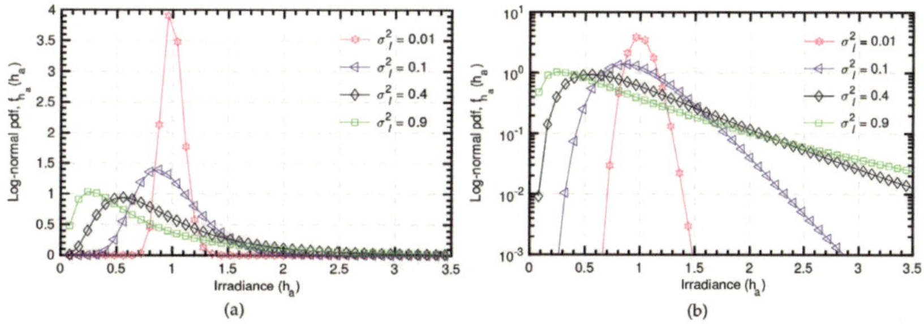

Figure 12. Log-normal pdf for different values of log-irradiance variance using (a) linear scale and (b) logarithmic scale.

We substitute different values of the log-irradiance variance, that is, $\sigma_l^2 \in$ (0.01, 0.1, 0.4, 0.9) in Eq. (11) into Eq. (10) in order to estimate the intensity fluctuation pdf for the LN distribution. The LN pdf plot can be easily achieved by using a linear scale; however, we also present results of a logarithmic scale implementation. This is to ensure uniformity in this chapter because we are going to use the logarithmic scale to present experimental results of irradiance fluctuation in Section 6. Moreover, logarithmic scale has the ability to respond to skewness toward the large values. So, it is more appropriate for illustrating how the measured samples would fit with the distributions at their tails. **Figure 12** shows the LN pdf for different values of log-irradiance variance. In **Figure 12a**, linear scale is employed whereas in **Figure 12b** we use logarithmic scale. It is crystal clear that detailed insight of the tails of the plots is presented in **Figure 12b**. Furthermore, it should be noted that as the value of σ_l^2 increases, the distribution is getting more and more tilted. This signifies the magnitude of the irradiance fluctuation of the system.

3.3.2. Gamma-gamma (ΓΓ) distribution

In most cases, in the strong turbulence regimes in which the LN distribution characterization is not valid, the ΓΓ distribution is normally employed in modeling the scintillation effects. Also, the ΓΓ model can be used in characterizing the fading gains from the weak to the strong turbulence scenarios. The pdf of h_a using the ΓΓ distribution is given by [3, 5, 35, 39]

$$f_{h_a}(h_a) = \frac{2(\alpha\beta)^{(\alpha+\beta)/2}}{\Gamma(\alpha)\Gamma(\beta)}(h_a)^{\frac{(\alpha+\beta)}{2}-1}K_{\alpha-\beta}(2\sqrt{\alpha\beta h_a}), \tag{18}$$

where $\Gamma(\cdot)$ represents the gamma function, $K_\nu(\cdot)$ is the modified Bessel function of the second kind of order ν, α and β are the effective number of large-scale and small-scale eddies of the scattering process, respectively. The parameters α and β are defined, respectively, for the plane wave as [3, 5, 35, 39]

$$\alpha = \left[\exp\left(\frac{0.49\sigma_R^2}{(1+1.11\sigma_R^{12/5})^{7/6}}\right)-1\right]^{-1}, \tag{19}$$

$$\beta = \left[\exp\left(\frac{0.51\sigma_R^2}{(1+0.69\sigma_R^{12/5})^{5/6}}\right)-1\right]^{-1}, \tag{20}$$

and for the spherical wave, they can be expressed as [35]

$$\alpha = \left[\exp\left(\frac{0.49\sigma_R^2}{(1 + 0.18d^2 + 0.56\sigma_R^{12/5})^{7/6}} \right) - 1 \right]^{-1}, \tag{21}$$

$$\beta = \left[\exp\left(\frac{0.51\sigma_R^2 (1 + 0.69\sigma_R^{12/5})^{-5/6}}{(1 + 0.9d^2 + 0.62d^2\sigma_R^{12/5})^{5/6}} \right) - 1 \right]^{-1}, \tag{22}$$

where $d \triangleq (kD^2/4L)^{1/2}$, D represents the diameter of the receiver aperture, and σ_R^2 denotes the Rytov variance which is a metric for the strength of the turbulence fluctuations. The σ_R^2 is defined for the plane and the spherical waves, respectively, as [5, 35, 39]

$$\sigma_R^2|_{\text{plane}} = 1.23\ C_n^2\ k^{7/6}\ L^{11/6}, \tag{23}$$

$$\sigma_R^2|_{\text{spherical}} = 0.492\ C_n^2\ k^{7/6}\ L^{11/6}, \tag{24}$$

Apart from different values of σ_l^2 being employed, that is, $\sigma_l^2 \in$ (4.0, 1.5, 0.5), we substitute different values of the effective number of large- and small-scale eddies of the scattering process, that is, α and $\beta \Rightarrow \alpha \in (4.3, 4.1, 6.0)$ and $\beta \in (1.3, 2.0, 4.4)$ in Eq. (19) and Eq. (20) into Eq. (18) to estimate the ΓΓ pdf. The results of both linear and logarithmic scales are depicted in **Figure 13**. **Figure 13** shows turbulence regimes that correspond to the weak, moderate, and strong atmospheric scenarios. The results show that an increase in the turbulence from the weak regime to the strong regime leads to corresponding increase in the distribution spreading.

The normalized variance of the irradiance also known as the scintillation index (σ_N^2) can be expressed in terms of σ_x^2 and eddies of the scattering process (α and β), respectively, as [35, 39]

$$\sigma_N^2 \triangleq \frac{\langle h_a^2 \rangle - \langle h_a \rangle^2}{\langle h_a \rangle^2} \tag{25}$$

$$= \frac{\langle h_a^2 \rangle}{\langle h_a \rangle^2} - 1 \tag{26}$$

$$= exp(4\sigma_x^2) - 1 \tag{27}$$

$$= 1/\alpha + 1/\beta + 1/(\alpha\beta). \tag{28}$$

3.3.3. *Málaga ($\mathcal{M}$)-distribution*

The $\mathcal{M}$-distribution is a generic model that is appropriate for defining the entire turbulent regimes. **Table 4** shows the means of generating the existing atmospheric turbulence models from the $\mathcal{M}$-distribution model [41]. The $\mathcal{M}$-distributed fading model is based on components such as U_L, U_S^C, and U_S^G which represent the line-of-sight (LOS) component, the scattered component by the eddies on the propagation axis that is coupled to the LOS contribution, and the scattered component to the receiver by the off-axis eddies, respectively. The average power of the LOS component as well as that of the total scatter components are $\Omega = E[|U_L|^2]$

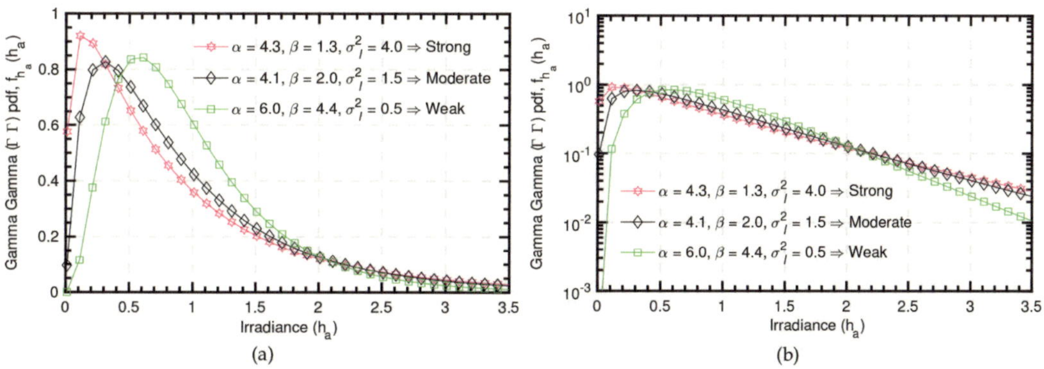

Figure 13. Gamma-gamma pdf for weak-strong turbulence regimes using (a) linear scale and (b) logarithmic scale.

Distribution model	Generation									
	ρ	Var[$\lvert U_L \rvert$]	Var[G]	X	Ω'	Var[X]	γ	Ω	α	β
Rice-Nakagami	0	0								
Gamma	0						0			
Homodyned K (HK)	0		0	γ						
Gamma-gamma	1				1		0			
Shadowed-Rician						0				
Log-normal	0	0					$\to 0$			
K	0							0		
Exponential	0	0						0	$\to \infty$	
Gamma-Rician										$\to \infty$

Adapted from Ref. [41].

Table 4. Approximations required for generating different distribution models from *M*-distribution model.

and $2b_0 = E[|U_S^C|^2 + |U_S^G|^2]$, respectively. In addition, the average power of the coupled-to-LOS scattering component and that of the scattering component received by off-axis eddies are $E[|U_S^C|^2] = 2\rho b_0$ and $E[|U_S^G|^2 = (1-\rho)2b_0]$, respectively. The parameter $0 \le \rho \le 1$ represents the amount of scattering power coupled to the LOS component [41, 42]. The pdf of the optical channel gain h_a for the M-distribution can be written as [41, 42]

$$f_{h_a}(h_a) = A\sum_{k=1}^{\beta} a_k h_a^{\frac{\alpha+k}{2}-1} K_{\alpha-k}\left(2\sqrt{\frac{\alpha\beta h_a}{\mu\beta + \Omega'}}\right) \tag{29}$$

where $\mu = E[|U_S^C|]^2 = (1-\rho)2b_0$, $\Omega' = \Omega + 2\rho b_0 + 2\sqrt{2\rho b_0 \Omega} cos(\varphi_A - \varphi_B)$, φ_A and φ_B represent the deterministic phases of the LOS and the coupled-to-LOS component, respectively; α denotes a positive parameter that depends on the effective number of large-scale cells of the scattering process, β is a natural number which represents the amount of fading parameter, $K_v(\cdot)$ represents the modified Bessel function of the second kind with order v, A and a_k can be expressed, respectively, as [41, 42]

$$A \triangleq \frac{2\alpha^{\frac{\alpha}{2}}}{\mu^{1+\frac{\alpha}{2}}\Gamma(\alpha)}\left(\frac{\mu\beta}{\mu\beta+\Omega'}\right)^{\beta+\frac{\alpha}{2}} \tag{30}$$

$$a_k \triangleq \begin{pmatrix} \beta & -1 \\ k & -1 \end{pmatrix} \frac{(\mu\beta+\Omega')^{1-\frac{k}{2}}}{(k-1)!}\left(\frac{\Omega'}{\mu}\right)^{k-1}\left(\frac{\alpha}{\beta}\right)^{\frac{k}{2}} \tag{31}$$

3.4. Combined attenuation statistics

The pdf of $h = h_\ell h_a h_p$ that constitutes the aforementioned factors of the propagation channel can be defined as [37, 38]

$$f_h(h; w_z) = \int f_{h|h_a}(h|h_a) f_{h_a}(h_a) dh_a \tag{32}$$

where $f_{h|h_a}(h|h_a)$ is the conditional probability given a turbulence state h_a and its distribution can be written as [37, 38]

$$f_{h|h_a}(h|h_a) = \frac{1}{h_a h_\ell} f_{h_p}\left(\frac{h}{h_a h_\ell}\right) = \frac{\gamma^2}{A_0^{\gamma^2} h_a h_\ell}\left(\frac{h}{h_a h_\ell}\right)^{\gamma^2-1}, \qquad 0 \le h \le A_0 h_a h_\ell. \tag{33}$$

Therefore, $f_h(h; w_z)$ can be written as [37]

$$f_h(h; w_z) = \frac{\gamma^2}{(A_0 h_\ell)^{\gamma^2}} h^{\gamma^2-1} \int_{h/A_0 h_\ell}^{\infty} h_a^{-\gamma^2} f_{h_a}(h_a) dh_a. \tag{34}$$

4. Performance analysis

4.1. BER

The channel state distribution $f_h(h; w_z)$ can be calculated by employing an appropriate model for atmospheric turbulence regimes in Eq. (25) as follows:

1. **For a weak turbulence** ($\sigma_R^2 < 0.3$)

   In this regime, $f_{h_a}(h_a)$ follows LN distribution, so, $f_h(h; w_z)$ can be expressed as [37]

$$f_h(h; w_z) = \frac{\gamma^2}{2(A_0 h_\ell)^{\gamma^2}} h^{\gamma^2-1} \times \text{erfc}\left(\frac{\ln\left(\frac{h}{A_0 h_\ell}\right)+\mu}{\sqrt{8}\sigma_x}\right) e^{\left(2\sigma_x^2\gamma^2(1+\gamma^2)\right)}. \tag{35}$$

2. **For a strong turbulence**

   This regime is characterized by the $\Gamma\Gamma$ distribution, so, $f_h(h; w_z)$ can be written as [37, 38]

$$f_h(h; w_z) = \frac{2\gamma^2(\alpha\beta)^{(\alpha+\beta)/2}}{(A_0 h_\ell)^{\gamma^2}\Gamma(\alpha)\Gamma(\beta)} h^{\gamma^2-1} \times \int_{h/A_0 h_\ell}^{\infty} h_{\mathrm{a}}^{(\alpha+\beta)/2-1-\gamma^2} \mathrm{K}_{\alpha-\beta}(2\sqrt{\alpha\beta\, h_{\mathrm{a}}}) dh_{\mathrm{a}}. \tag{36}$$

where $\mathrm{K}_\nu(\cdot)$ can be expressed in terms of the Meijer's G-function $G_{p,q}^{m,n}[\cdot]$ as [35]

$$\mathrm{K}_\nu(x) = \frac{1}{2} G_{0,2}^{2,0}\left[\frac{x^2}{4}\middle|_{(\nu/2),-(\nu/2)}^{-}\right]. \tag{37}$$

So, from Eq. (28), $\mathrm{K}_{\alpha-\beta}(2\sqrt{\alpha\beta h_{\mathrm{a}}})$ can be expressed as [35]

$$\mathrm{K}_{\alpha-\beta}(2\sqrt{\alpha\beta h_{\mathrm{a}}}) = \left(\frac{1}{2}\right) \mathrm{G}_{0,2}^{2,0}\left[\alpha\beta h_{\mathrm{a}}\middle|_{\frac{\alpha-\beta}{2},\frac{\beta-\alpha}{2}}^{-}\right], \tag{38}$$

Therefore, $f_h(h)$ can be written as [38]

$$f_h(h) = \frac{\alpha\beta\gamma^2}{A_0 h_l \Gamma(\alpha)\Gamma(\beta)} \left(\frac{\alpha\beta h}{A_0 h_l}\right)^{(\alpha+\beta/2)-1} \times G_{1,3}^{3,0}\left[\frac{\alpha\beta}{A_0 h_l} h \middle|_{-\frac{\alpha+\beta}{2}+\gamma^2,\frac{\alpha-\beta}{2},\frac{\beta-\alpha}{2}}^{1-\frac{\alpha+\beta}{2}+\gamma^2}\right]. \tag{39}$$

The average BER, $P(e)$, in terms of $P_h(h)$ can be expressed as [38]

$$P(e) = \int_0^\infty P(e|h) f_h(h) dh. \tag{40}$$

Figure 14 shows the average BER in terms of the average SNR for BPSK over different values of turbulence strength. In this analysis, only the atmospheric turbulence effect is assumed

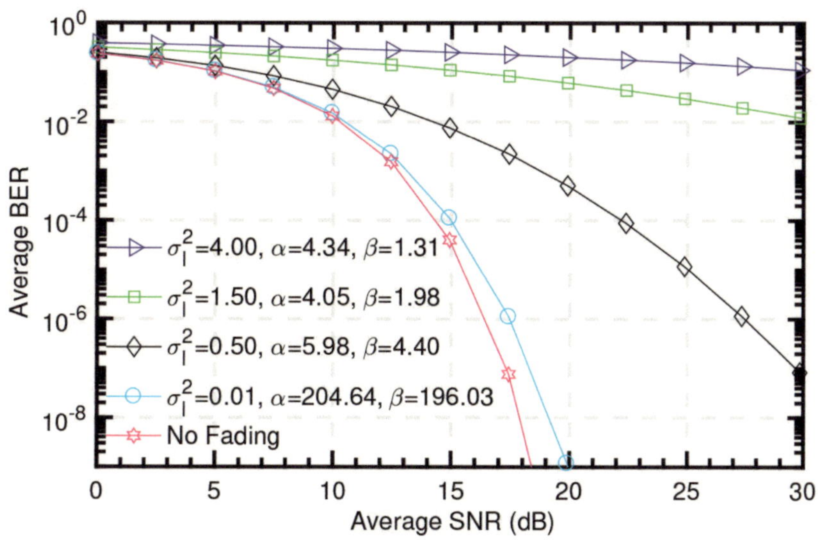

Figure 14. Average BER versus SNR for BPSK under different turbulence conditions.

when Eq. (40) is considered. The atmospheric turbulence parameters $\alpha \in$ (4.34, 4.05, 5.98, 204.64), $\beta \in$ (1.31, 1.98, 4.40, 196.03), and $\sigma_l^2 \in$ (4.00, 1.50, 0.50, 0.01), which represent weak to strong turbulence conditions, are employed in order to estimate the system BER performance. It is observed that the SNR required to achieve a specific BER increases with an increase in the atmospheric turbulence strength. For instance to achieve a BER of 10^{-6} in a channel with $\sigma_l^2 = 0.01$, the required SNR is about 18 dB; however, for fading strength of $\sigma_l^2 = 0.50$, the required SNR increases to 28 dB. Furthermore, at this BER, for $\sigma_l^2 = 0.01$ and $\sigma_l^2 = 0.50$ additional 2 and 12 dB, respectively, are required compared to the ideal channel in which there is no turbulence. This shows that the BER increases as turbulence strength becomes stronger.

4.2. Ergodic channel capacity

The channel capacity is one of the main performance metrics in the design of FSO systems that needs significant attention. The capacity of a multiple-input-multiple-output (MIMO) FSO system with M lasers and N photodetectors in bits/s/Hz can be expressed as [34]

$$C = \log_2\left[det\left(\boldsymbol{I}_M + \frac{\gamma_{\text{inst}}}{M}\boldsymbol{R}\right)\right], \tag{41}$$

where $\boldsymbol{R} = \begin{cases} \boldsymbol{H}\boldsymbol{H}^\dagger \text{ if } N < M \\ \boldsymbol{H}^\dagger\boldsymbol{H} \text{ if } N \leq M \end{cases}$, $\boldsymbol{H}$ denotes an $N \times M$ channel state matrix, $(\cdot)^\dagger$ corresponds to the Hermitian transpose, $\boldsymbol{I}_M$ is an $M \times M$ identity matrix, and $\gamma_{\text{inst}} = \eta_e^2 \text{h}^2/N_0$ denotes the instantaneous electrical SNR whose average $\xi_a = \eta_e^2 \mathbb{E}\langle \text{h} \rangle^2/N_0$.

It is worth noting that there are a number of viable approaches for the estimation of ergodic capacity of the MIMO FSO link. One of such is the numerical integration approach.

4.2.1. Numerical integration approach

The ergodic capacity, C_{erg}, of MIMO FSO link can be defined by the expected value of the instantaneous mutual information C, between the transmit and receive apertures. Hence, the C_{erg} of FSO system is a random variable and a function of SNR. The C_{erg} can be expressed as [34, 35, 43]

$$C_{\text{erg}} \triangleq \mathbb{E}\langle C \rangle = \int_0^\infty \log_2(1 + \gamma_{\text{inst}}) f_{\gamma_{\text{inst}}}(\gamma_{\text{inst}}) d\gamma_{\text{inst}}, \tag{42}$$

where $\mathbb{E}(\cdot)$ denotes the expectation operator and $f_{\gamma_{\text{inst}}}(\gamma_{\text{inst}})$ is the pdf of γ_{inst}.

It should be noted that most of the models for ergodic capacity evaluation are based on numerical integration. However, integration-based approach requires comparatively more computational time. This is even more challenging in the strong turbulence regimes analysis. This consequence can be attributed to the Bessel function that is usually expressed in terms of the Meijer's *G*-function for easier evaluation [34]. In Refs. [34] and [35], computational-efficient approaches are presented in order to reduce the associated high computational time of the integration-based approach. This enables faster performance evaluation over a wide range of SNR.

4.2.2. Power series approach

Considering the power series representation, the ergodic channel capacity can be expressed as [35]

$$\langle C\rangle = f(\mathcal{X}) = \psi\mathcal{A}(a_0 + a_1\mathcal{X} + a_2\mathcal{X}^2 + a_3\mathcal{X}^3), \tag{43}$$

where $\psi = \pi/4$ and $\mathcal{A} = \min\{M, N\}$ represent the minimum number of transmit or receive apertures, $\mathcal{X} = k_1\gamma$, k_1 is a coefficient whose value is one and the unit is in b/s/Hz, and a_n represents the coefficient of the nth term given by [35]

$$a_0 = 0.13\pi k_1\mathfrak{R}, \tag{44}$$

$$a_1 = 0.66\mathfrak{R}, \tag{45}$$

$$a_2 = 1.45 \times 10^{-3}\mathfrak{R}, \tag{46}$$

$$a_3 = -1.73 \times 10^{-5}\mathfrak{R}, \tag{47}$$

where $\mathfrak{R} = \pi\exp(\sigma_R^2 - \sigma_N^2)$.

4.2.3. Spatial interpolation lookup approach

The spatial approach is based on B-spline and Barycentric Lagrange interpolation lookup table (B^2LUT). With the B-spline interpolated LUT, the capacity of MIMO FSO system can be expressed as [34]

$$\begin{aligned} C\left(\mathrm{SNR}|_{(M,N,\sigma_N^2)}\right) &= \sum_{t=1}^{m_1}\sum_{r=1}^{m_2}\sum_{p=1}^{m_3}\omega(\mathrm{SNR}|_{\sigma_N^2}\mathfrak{I}_p, \cdots, \mathfrak{I}_{p+h}) \\ &\times \psi(\mathrm{SNR}|_N\mathfrak{D}_r, \cdots, \mathfrak{D}_{r+\ell})\phi(\mathrm{SNR}|_M\mathfrak{S}_t, \cdots, \mathfrak{S}_{t+s})\mathfrak{M}_{t,r,p}, \end{aligned} \tag{48}$$

where $\mathfrak{I} = (\mathfrak{I}_p, \cdots, \mathfrak{I}_{p+h})$, $\mathfrak{D} = (\mathfrak{D}_r, \cdots, \mathfrak{D}_{r+\ell})$, and $\mathfrak{S} = (\mathfrak{S}_t, \cdots, \mathfrak{S}_{t+s})$ are the knot sequences, $(\mathfrak{M}_{t,r,p} : t = 1, \cdots, m_1; r = 1, \cdots, m_2; p = 1, \cdots, m_3)$ are coefficients array, and $\omega(\cdot), \psi(\cdot), \phi(\cdot)$ are the univariate piecewise B-spline for M, N, and σ_N^2, respectively.

Another main figure of merit for characterizing the communication link performance is the achievable average (ergodic) channel capacity. **Figure 15** shows the average channel capacity of FSO link as a function of average electrical SNR for different values of turbulence strength. The atmospheric turbulence parameters $\sigma_N^2 \in$ (0.0120, 0.1486, 0.2078) result in $C_N^2 \in$ $(6.03 \times 10^{-16}m^{-2/3}, 7.62 \times 10^{-15}m^{-2/3}, 1.09 \times 10^{-14}m^{-2/3})$, which correspond to the weak, moderate, and strong turbulence conditions, respectively. Moreover, the capacity of the nonturbulent channel condition (no fading) is presented for benchmarking. Obviously, the ergodic capacity of FSO link significantly depends on the atmospheric turbulence strength.

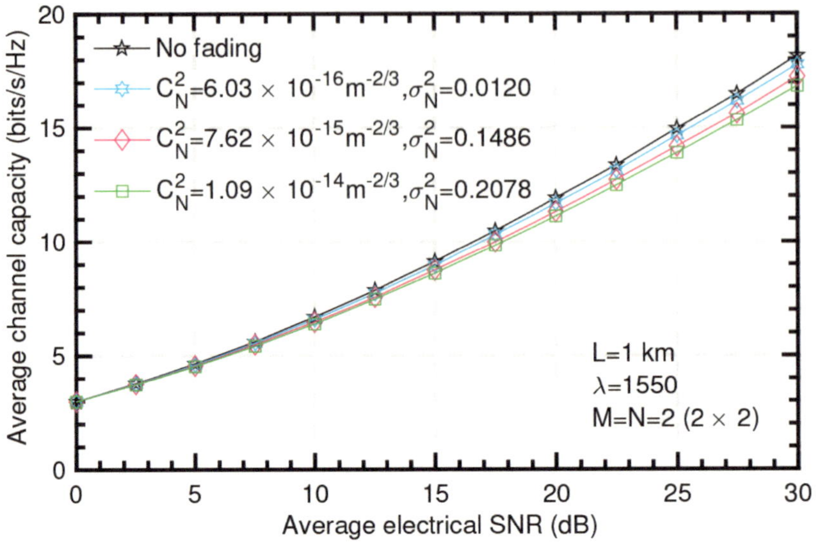

Figure 15. Average channel capacity of FSO link versus average electrical SNR.

Comparatively, the ergodic channel capacity for weak turbulence condition is considerably more than in the cases of moderate and strong turbulence conditions. Moreover, the channel condition with no fading offers the highest capacity. This shows that the atmospheric turbulence-induced fading results in severe impairment on the FSO link performance. This can eventually result in recurrent link failures. Consequently, optical wireless technologies are not as reliable as the conventional optical fiber technologies. In order to address the challenges, innovative technologies can be employed to enhance the system performance.

5. Technologies for performance enhancement

FSO is a promising optical technology that can be employed for different application. However, the tradeoff between the required high data rates and the limitations of atmospheric channel is the major challenge for reliable implementation of FSO technologies in the access networks [35]. Therefore, the problems inhibit the FSO system from being an effective and reliable standalone fronthaul technology. In this section, schemes like hybrid RF/FSO and relay-assisted transmission technologies that can be implemented to enhance the performance of FSO technology in the access networks are presented.

5.1. Hybrid RF/FSO technology

It should be noted that RF wireless technologies that operate above approximately 10-GHz frequencies are adversely affected by rain, whereas fog has insignificant effect on them. On the other hand, FSO systems are highly susceptible to fog, whereas the effect of rain on them is negligible. Therefore, it is of high importance to improve the link reliability to alleviate the adverse effects of the meteorological and weather conditions. An attractive way of addressing the challenge is a simultaneous employment of an RF link and the FSO link for transmission. It is remarkable that fog and rain rarely occur concurrently in nature. Consequently, the two links can function in a complementary way. This concept influences the hybrid RF/FSO scheme. The RF/FSO is a hybrid scheme that combines the benefits of the inherent high transmission capacity of optical technologies and the ease of deployment of wireless links. Moreover, the idea of hybrid RF/FSO system is to concurrently attend to the related drawbacks and take advantages of both technologies. This will help in the reliable transmission of heterogeneous wireless services [8, 24].

In a hybrid RF/FSO technology, there are two parallel links between the transmitter and the receiver. Moreover, subject to deployment scenario and application, both parallel links of the hybrid technology have the capability to transmit data. Nevertheless, based on the weather conditions as well as the EMI levels, either of the links can be used for data transmission [44]. For instance, under adverse atmospheric condition (i.e., fog), the hybrid RF/FSO scheme ensures that the RF link serves as a back-up in case of FSO link outage. However, the resultant data rate of the RF link is less than that of the actual FSO link [8, 24]

5.2. Relay-assisted FSO transmission

A realistic approach for turbulence-induced fading mitigation is spatial diversity technique. In this technique, multiple transmit/receive apertures are employed in order to create and exploit additional degrees of freedom in the spatial domain. The spatial diversity is an appealing technique for fading mitigation because of its typical redundancy. However, the utilization of multiple apertures presents different challenges such as an increase in the system complexity as well as cost. In addition, the distance between the apertures has to be large enough in order to inhibit detrimental effects of spatial correlation. A simplified way of implementing spatial diversity is the dual-hop relaying which has been considerably employed in the RF communication systems. The dual-hop relaying implementation helps substantially in extending the network coverage area as well as improving the quality of the receive signal [45].

Furthermore, the concept of relay-assisted transmission is based on creating a virtual multiple-aperture system in order to realize advantages of MIMO techniques. This is achieved by exploiting both RF and FSO characteristics in order to have an efficient system in a real-life situation. Additionally, a relay-assisted transmission is also known as a mixed RF/FSO dual-hop communication scheme. The dual-hop entails the links from the source to the relay which are RF links and the links between the relay and the destination which are FSO links. In essence, RF transmission is utilized at one hop and FSO transmission is employed at the other. It is remarkable that, in principle, the mixed RF/FSO dual-hop relay scheme is comparatively different from the hybrid RF/FSO technology. In the latter, parallel RF and FSO links are normally used for the same path [45]. Furthermore, in the mixed RF/FSO dual-hop scheme, the main purpose of the FSO link is to enable the RF users to communicate with the backbone network. This helps bridging the connectivity gap between the backbone and the last-mile access networks [42, 46].

The mixed RF/FSO dual-hop model efficiently addresses the last-mile transmission bottleneck of the system. This is achieved by enabling multiplexing of multiple users with RF capabilities as well as their aggregation in a particular high-speed FSO link so as to exploit the inherent optical capacity [42, 45]. Moreover, this implementation stalls any form of interference due to the fact that RF and FSO operate on completely different frequency bands. Therefore, the mixed RF/FSO dual-hop model offers better performance compared to the traditional RF/RF transmission system [45, 46]. In a mixed RF/FSO system, the source comprises multiple RF users, each equipped with an antenna. Furthermore, at the destination, there is an FSO detector that is equipped with an aperture. In addition, the source and the destination are connected by a relay that is usually mounted on a high platform. The relay node performs RF to FSO conversion. Also, the relay has a receive antenna and a transmit aperture that are assigned for the RF signal reception and optical signal transmission, respectively.

6. Channel measurement and characterization

In this section, the performance of an FSO link subjected to a real atmospheric turbulence condition is investigated experimentally. The σ_N^2 is measured from the channel samples obtained so as to determine the degree of atmospheric turbulence and the subsequent effects on the FSO link quality. Furthermore, the C_n^2 can be calculated as explained in Section 3.

6.1. Experimental setup

The experimental setup shown in **Figure 16** is employed in the channel measurement. The setup consists of a point-to-point FSO link that is based on IM/DD technique. The pattern generator uses a pseudorandom binary sequence (PRBS) of length $2^{23} - 1$ bits to generate a 10-Gb/s non-return-to-zero (NRZ) signal. Also, the produced electrical signal is then injected into a JDSU Integrated Laser Mach Zehnder (ILMZ) drives at 1548.51-nm wavelength. A standard single-mode fiber (SSMF) is used to convey the optical output signal launched from the laser to a 3-mm diameter collimator. The optical power at the input of collimator is set to 0 dBm. The collimated laser beam is subsequently transmitted over the FSO channel with a round trip length of 54 m. **Figure 16** inset depicts an outdoor FSO setup employed in the measurement. The overall transmission distance is achieved when the optical signal from collimator 1 passes

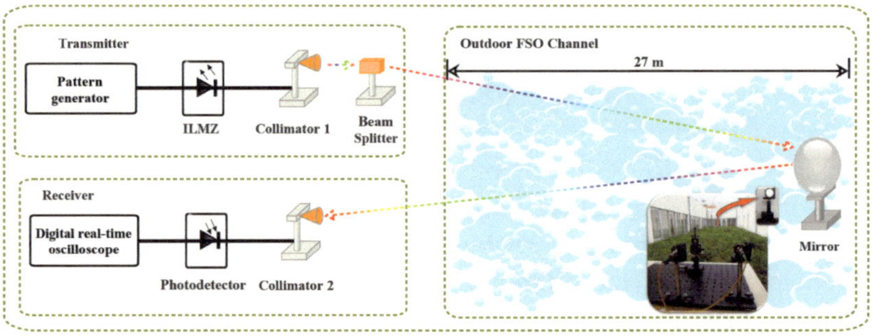

Figure 16. Experimental setup. The inset presents the picture of outdoor FSO setup, ILMZ: Integrated Laser Mach Zehnder.

through the FSO channel to the mirror with a beam diameter of about 2 cm located at the other side of the link that is 27 m long, and then reflected back to the collimator 2 at the receiver with approximately its initial diameter. The collimator is made of a concave mirror, in order to lessen beam scattering and considerably maximize the power transfer. The converged received optical signal at the receiver then focuses on the laser collimator which is coupled to the photodetector by the SSMF. The resulting optical signal is then converted into electrical signal using a 10-Gb/s PIN photodiode. The PIN is followed by a real-time sampling oscilloscope (Tektronix: DPO72004B) with a sample rate of 50 GS/s.

6.2. Experimental results

The results of FSO channel samples collected from November 9 to 20, 2015, for characterization are presented. The data obtained on November 12, 2015, at 01:45 pm and 09:30 pm are analyzed in this work. The recorded weather conditions are as follows:

1. Scenario 1: 01:45 pm

   Temperature, 22°C; wind, 6 mph; humidity, 69%; pressure, 1031 mb; visibility, 10 km; precipitation, 0 mm; and rain rate, 0%.

2. Scenario 2: 09:30 pm

   Temperature, 17°C; wind, 4 mph; humidity, 80%; pressure, 1030 mb; visibility, 9 km; precipitation, 0 mm; and rain rate, 0%.

The FSO channel measurement is realized by injecting an unmodulated optical wave that emanates from the laser into the FSO channel. The FSO channel brings about a path loss with

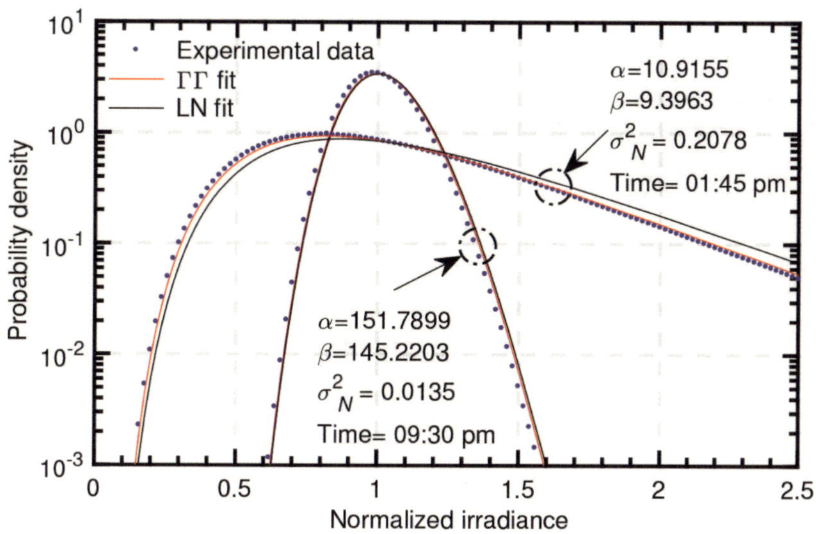

Figure 17. Histogram of normalized irradiance with log-normal and gamma-gamma fits under different scintillation index values.

a mean value of 9 dB, for the setup and the atmospheric conditions. Using statistical means, the resultant signal detected and received by the real-time oscilloscope is analyzed offline using MATLAB®.

The characterization of the refractive-index structure parameter C_n^2 is achieved by fitting the nearest LN and the ΓΓ pdf curves to the pdf of the received data. The fittings are presented in **Figure 17**. The scintillation index σ_N^2 is measured for the two scenarios considered. The values obtained are 0.0135 and 0.2078 for 09:30 pm and 01:45 pm, respectively. For the first scenario with $\sigma_N^2 = 0.0135$, the LN and the ΓΓ fit very well with the measured channel samples σ_N^2. However, when $\sigma_N^2 = 0.2078$, the LN fitting is loose and unable to give an accurate result for the fading model, whereas the ΓΓ fitting still maintains a relatively better result for the fading model. This result shows that the LN model is unsuitable for the strong atmospheric fading characterization.

Furthermore, the estimated values of the refractive-index structure parameters, C_n^2, are $6.7807 \times 10^{-16} m^{-2/3}$ ($\sigma_N^2 = 0.0135$) and $1.0864 \times 10^{-14} m^{-2/3}$ ($\sigma_N^2 = 0.2078$). Therefore, the first ($\sigma_N^2 = 0.2078$) and the second ($\sigma_N^2 = 0.0135$) scenarios correspond to the strong and weak turbulence regimes, respectively.

7. Real-time coherent PON OWC based on dual polarization for the mobile backhaul/fronthaul

In this section, proof-of-concept gigabit-capable long-reach coherent PON and OWC systems with the ability to support different applications over a shared optical fiber infrastructure are experimentally implemented. This is in an effort to demonstrate the FSO application in certain areas in the mobile cellular systems in which physical connections by means of optical fiber cables are impractical or in rural area that lacks fiber infrastructure. Moreover, this is achieved by a reconfigurable real-time DSP reception of a dual-polarization-quadrature phase shift keying (DP-QPSK) signal over the SSMF and FSO systems. It is worth noting that the system is validated by a commercial field-programmable gate array (FPGA) in order to have an open system whose components and protocols conform to standards independent of a particular equipment vendor. In this analysis, we study signal transmission and reception over 100 km of SSMF as well as over a hybrid 100 km of SSMF plus a 54-m outdoor FSO link. We are able to establish the lowest sampling rate that is necessary for digital coherent PON by employing four 1.25 Gsa/s ADCs with an electrical front-end receiver that offers just 1-GHz analog bandwidth. This is realized by implementing a phase and polarization diversity coherent receiver in conjunction with the DP-QPSK modulation formats. This technique is of high importance in order to relax the required electrical digital units at the optical network unit (ONU) toward the RF rates. This scheme also helps in realizing the anticipated data rate for the next-generation coherent optical access networks for the 5G Mobile wireless networks.

7.1. Experimental setup

The experimental setup depicted in **Figure 18a** is used to validate the performance of a PON system with hybrid fiber and FSO link using DP signals. It is worth mentioning that only the

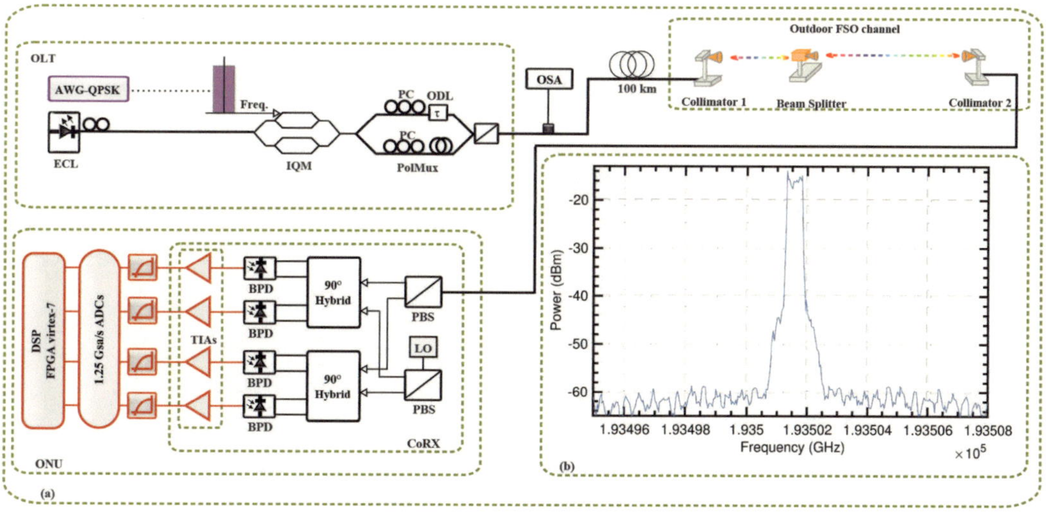

Figure 18. (a) Experimental setup for 20×625 Mbaud DP-QPSK signal; (b) Overall spectrum (PBS: polarization beam splitter; BPD: balanced photodetector; CoRX: coherent receiver).

receiver DSP of the setup is estimated in real time. At the OLT, the light from an external cavity laser (ECL) (<100 kHz linewidth) is injected into an IQ modulator (IQM). The wavelength λ is centered at ~1549 nm. The IQM is driven by a 65-Gsa/s arbitrary waveform generator (AWG) that generates 625 Mbaud signals from a $2^{12} - 1$ pseudo random bit sequence (PRBS). The subsequent signal is then digitally filtered using a raised-cosine (RC) filter with a 0.1-roll-off factor and 32-taps FIR resolution in addition to a simple 3-taps FIR pre-emphasis subsystem. The employed modulation format is the differential DP-QPSK, providing 2.5 Gb/s per end user. The obtained spectrum is shown in **Figure 18b**.

In an effort to emulate the DP system, this signal is divided into two using an optical splitter. Then, we applied a delay of 12 symbols to one of them for an effective decorrelation purposes. Afterward, both polarized signals are multiplexed orthogonally once again using a polarization beam combiner (PBC). This implementation results in dual polarization of the signal. The optical power is managed using variable optical attenuators (VOAs). This subsequent signal is then propagated over 100 km of an SSMF and 54-m FSO. At the receiver side, the signal is coherently detected using a 4×90° optical hybrid by means of a free-running ECL LO with about 100-kHz linewidth tuned to the center channel, λ. The optical signal is converted to the electrical domain using four balanced detectors (BDs) and then amplified by transimpedance amplifiers (TIAs). This results in the in-phase and quadrature components of each polarization. It is worth noting that only an output of the TIA is used. The signal is then filtered using a 1-GHz low-pass filter and sampled by four 8-bit 1.25 Gsa/s ADCs. The digitalized signal is conveyed to a Virtex-7 FPGA, where the entire post-detection 8-bit DSP in real time is implemented. The applied DSP is based on [47]. The bit-error rate (BER) is calculated in real time by bit error counting, averaged between the two polarizations.

In addition, we consider the possibility of an outdoor FSO communication link as part of the system. The employed FSO link setup in this study is similar to that in **Figure 16** that we have

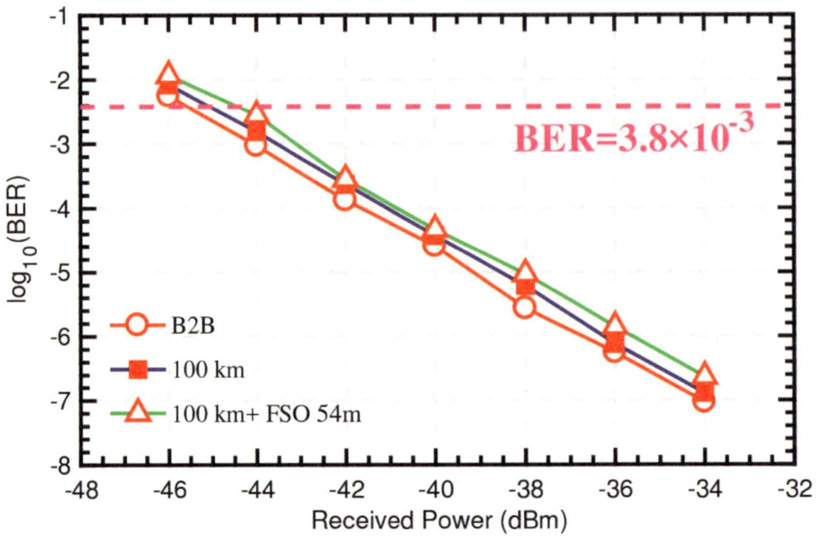

Figure 19. Receiver sensitivity in terms of BER measured for DP-QPSK signals.

discussed in Section 6. The outdoor FSO link experiences a total loss of ~8–9 dBm. At the collimator, the received signal is guided to an integrated phase- and polarization-diverse coherent receiver.

7.2. Experimental results

Figure 19 illustrates receiver sensitivity in terms of BER measured for the DP-QPSK signal. The figure presents results for the back-to-back (B2B) and 100 km of fiber scenarios as well as 100-km plus 54-m FSO. The considered BER limit of 3.8×10^{-3} corresponds to the 7% hard-decision forward error correction (HD-FEC). As shown in **Figure 19**, there is no significant penalty between 100 and 100-km plus FSO in real-time results.

8. Conclusions

In this chapter, we have presented various opportunities of using optical wireless communication technology for addressing the last-mile transmission bottleneck of the fixed/mobile network. We have also discussed various challenges of the OWC and presented different solutions in order to make OWC an efficient technology. In the proof-of-concept experiment, we study the transmission capabilities of a PON based on DP signal in terms of receiver sensitivity. This is implemented with the real-time ONU receiver that is emulated by a commercial FPGA. This helps in facilitating an open system and hence enables interoperability, portability, and open software standards. The transmissions over 100 km of SSMF as well as over a hybrid 100 km of SSMF plus a 54-m outdoor FSO link are successfully realized considering 625 Mbaud DP-QPSK channel.

http://dx.doi.org/10.5772/intechopen.69113

Acknowledgements

This work is supported by the European Structural Investment Funds (ESIF), through the Operational Competitiveness and Internationalization Programme (COMPETE 2020) under FutPON project [Nr. 003145 (POCI-01 -0247-FEDER-003145)]. Also, it is funded by the Fundação para a Ciência e a Tecnologia under the grants PD/BD/52590/2014 and FRH/BPD/110889/2015.

Author details

Isiaka Alimi*, Ali Shahpari, Artur Sousa, Ricardo Ferreira, Paulo Monteiro and António Teixeira

*Address all correspondence to: iaalimi@ua.pt

Instituto de Telecomunicações, Department of Electronics, Telecommunications and Informatics, Universidade de Aveiro, Aveiro, Portugal

References

[1] Yu C, Yu L, Wu Y, He Y, Lu Q. Uplink scheduling and link adaptation for narrowband internet of things systems. IEEE Access, 2017;**5**:1724–1734

[2] Ejaz W, Anpalagan A, Imran MA, Jo M, Naeem M, Qaisar SB, Wang W. Internet of things (IoT) in 5G wireless communications. IEEE Access. 2016;**4**:10310–10314

[3] Alimi I, Shahpari A, Ribeiro V, Kumar N, Monteiro P, Teixeira A. Optical wireless communication for future broadband access networks. In: 2016 21st European Conference on Networks and Optical Communications (NOC); IEEE; Lisbon, Portugal. Jun 2016. pp. 124–128

[4] Parca G, Tavares A, Shahpari A, Teixeira A, Carrozzo V, Beleffi GT. FSO for broadband multi service delivery in future networks. In: 2013 2nd International Workshop on Optical Wireless Communications (IWOW); IEEE; Newcastle upon Tyne, UK. Oct 2013. pp. 67–70

[5] Ghassemlooy Z, Popoola W, Rajbhandari S. Optical wireless Communications: System and Channel Modelling with MATLAB®. Taylor & Francis; London, New York: 2012

[6] Ghassemlooy Z, Amon S, Uysal M, Xu Z, Cheng J. Emerging optical wireless communications-advances and challenges. IEEE Journal on Selected Areas in Communications. Sep 2015;**33** (9):1738–1749

[7] Alkholidi AG, Altowij KS. Free Space Optical Communications—Theory and Practices, Contemporary Issues in Wireless Communications. InTech; 2014. pp. 159–212

[8] Khalighi MA, Uysal M. Survey on free space optical communication: A communication theory perspective. IEEE Communications Surveys Tutorials. 2014;**16**(4):2231–2258

[9] Bloom S, Korevaar E, Schuster J, Willebrand H. Understanding the performance of free-space optics [Invited]. Journal of Optical Network. Jun 2003;**2**(6):178–200

[10] Kaushal H, Kaddoum G. Optical communication in space: Challenges and mitigation techniques. IEEE Communications Surveys Tutorials. 2017;**19**(1)57–96

[11] Zeng Z, Fu S, Zhang H, Dong Y, Cheng J. A survey of underwater optical wireless communications. IEEE Communications Surveys Tutorials. 2017;**19**(1):204–238

[12] Pompili D, Akyildiz IF. Overview of networking protocols for underwater wireless communications. IEEE Communications Magazine. Jan 2009;**47**(1):97–102

[13] Sevincer A, Bhattarai A, Bilgi M, Yuksel M, Pala N. FIGHTNETs: Smart FIGHTing and mobile optical wireless NETworks—A survey. IEEE Communications Surveys Tutorials. Fourth 2013;**15**(4):1620–1641

[14] Rajagopal S, Roberts RD, Pirn SK. IEEE 802.15.7 visible light communication: Modulation schemes and dimming support. IEEE Communications Magazine. Mar 2012;**50**(3):72–82

[15] Ying K, Yu Z, Baxley RJ, Qian H, Chang GK, Zhou GT. Nonlinear distortion mitigation in visible light communications. IEEE Wireless Communications. Apr 2015;**22**(2):36–45

[16] Yang F, Gao J. Dimming control scheme with high power and spectrum efficiency for visible light communications. IEEE Photonics Journal. Feb 2017;**9**(1):1–12

[17] Pinterest. Color wavelength frequency. Feb 2017. [Online]. Available from: https://www.pinterest.com/pin/408983209884752740/.

[18] Wang M, Wu J, Yu W, Wang H, Fi J, Shi J, Fuo C. Efficient coding modulation and seamless rate adaptation for visible light communications. IEEE Wireless Communications. Apr 2015;**22**(2):86–93

[19] ANDY. Visible Light Communication—VLC & PUREVLC™. February 2017 [Online]. Available from: http://andy96877.blogspot.pt/p/visible-light-communication-vlc-is-data.html

[20] Jan SU, Fee YD, Koo I. Comparative analysis of DIPPM scheme for visible light communications. In: 2015 International Conference on Emerging Technologies (ICET); IEEE; Peshawar, Pakistan. Dec 2015. pp. 1–5

[21] Tsouri GR, Zambito SR, Venkataraman J. On the benefits of creeping wave antennas in reducing interference between neighboring wireless body area networks. IEEE Transactions on Biomedical Circuits and Systems. Feb 2017;**11**(1):153–160

[22] Zhang R, Moungla H, Yu J, Mehaoua A. Medium access for concurrent traffic in wireless body area networks: Protocol design and analysis. IEEE Transactions on Vehicular Technology. Mar 2017;**66**(3):2586–2599

[23] Cahyadi WA, Jeong TI, Kim YH, Chung YH, Adiono T. Patient monitoring using visible light uplink data transmission. In: 2015 International Symposium on Intelligent Signal Processing and Communication Systems (ISPACS); IEEE; Nusa Dua, Indonesia. Nov 2015. pp. 431–434

[24] Uysal M, Capsoni C, Ghassemlooy Z, Boucouvalas A, Udvary E. Optical wireless communications: An emerging technology. Signals and Communication Technology. Springer International Publishing; Switzerland: 2016

[25] Baister G, Gatenby PV. Pointing, acquisition and tracking for optical space communications. Electronics Communication Engineering Journal. Dec 1994;**6**(6):271–280

[26] Toyoshima M. Trends in satellite communications and the role of optical free-space communications [Invited]. Journal of Optical Networking. Jun 2005;**4**(6):300–311

[27] D'Amico M, Feva A, Micheli B. Free-space optics communication systems: First results from a pilot field-trial in the surrounding area of Milan, Italy. IEEE Microwave and Wireless Components Letters. Aug 2003;**13**(8)505–307

[28] Song D-Y, Hurh Y-S, Cho J-W, Lim J-H, Fee D-W, Fee J-S, Chung Y. 4 × 10 Gb/s terrestrial optical free space transmission over 1.2 km using an EDFA preamplifier with 100 GHz channel spacing. Optics Express. Oct 2000;**7**(8):280–284

[29] Carlson RT, Paciorek S. Environmental Qualification and Field Test Results for the SONAbeam™ 155 and 622. Technical Report, fSONA Communications Corp. 2017; http://www.fsona.com/tech/white_papers/tech_qual-test.pdf

[30] Bandera P. Defining a Common Standard for Evaluating and Comparing Free-Space Optical Products. Technical Report, fSONA Communications Corp. 2017; http://www.fsona.com/tech/white_papers/WHTPAP-Generalized_Link_Margin.pdf

[31] LightPointe. Ultra-Low Latency Point-to-Point Wireless Bridge. White Paper, [Online]. Available from: http://nebula.wsimg.com/793e82b2beac48cb90c347bd86776d12?AccessKeyId=C1431E109BF92B03DF85&disposition=0&alloworigin=1.

[32] Wang Z, Zhong WD, Fu S, Fin C. Performance comparison of different modulation formats over free-space optical (FSO) turbulence links with space diversity reception technique. IEEE Photonics Journal. Dec 2009;**1**(6):277–285

[33] Navidpour SM, Uysal M, Kavehrad M. BER performance of free-space optical transmission with spatial diversity. IEEE Transactions on Wireless Communications. Aug 2007;**6**(8):2813–2819

[34] Alimi IA, Abdalla AM, Rodriguez J, Monteiro PP, Teixeira AF. Spatial interpolated lookup tables (FUTs) models for ergodic capacity of MIMO FSO systems. IEEE Photonics Technology Letters. Apr 2017;**29**(7)583–586

[35] Alimi I, Shahpari A, Ribeiro V, Sousa A, Monteiro P, Teixeira A. Channel characterization and empirical model for ergodic capacity of free-space optical communication link. Optics Communications. 2017;**390**:123–129

[36] Naboulsi MA, Sizun H, de Fornel F. Fog attenuation prediction for optical and infrared waves. Optical Engineering. 2004;**43**(2):519–329

[37] Farid AA, Hranilovic S. Outage capacity optimization for free-space optical links with pointing errors. Journal of Lightwave Technology. Jul 2007;**25**(7):1702–1710

[38] Sandalidis HG, Tsiftsis TA, Karagiannidis GK. Optical wireless communications with heterodyne detection over turbulence channels with pointing errors. Journal of Lightwave Technology. Oct 2009;**27**(20):4440–4445

[39] Andrews FC, Phillips RF. Laser Beam Propagation through Random Media. Press Monographs. SPIE Press; Bellingham, Washington USA. 2005

[40] Kiasaleh K. Performance of APD-based, PPM free-space optical communication systems in atmospheric turbulence. IEEE Transactions on Communications. Sep 2005;**53**(9):1455–1461

[41] Jurado-Navas A, Garrido-Balsells JM, Paris JF, Puerta-Notario A. A Unifying Statistical Model for Atmospheric Optical Scintillation. InTech; Rijeka, Croatia. 2011. pp. 181–206

[42] Yang F, Hasna MO, Gao X. Performance of mixed RF/FSO with variable gain over generalized atmospheric channels. IEEE Journal on Selected Areas in Communication. Sep 2015;**33**(9):1913–1924

[43] Peppas KP, Stassinakis AN, Topalis GK, Nistazakis HE, Tombras GS. Average capacity of optical wireless communication systems over I-K atmospheric turbulence channels. IEEE/OSA Journal of Optical Communications and Networking. Dec 2012;**4**(12):1026–1032

[44] Dahrouj H, Douik A, Rayal F, Al-Naffouri TY, Alouini MS. Cost-effective hybrid RF/FSO backhaul solution for next generation wireless systems. IEEE Wireless Communications. Oct 2015;**22**(5):98–104

[45] Zhang J, Dai F, Zhang Y, Wang Z. Unified performance analysis of mixed radio frequency/free-space optical dual-hop transmission systems. Journal of Lightwave Technology. Jun 2015;**33**(11):2286–2293

[46] Ansari IS, Yilmaz F, Alouini MS. Impact of pointing errors on the performance of mixed RF/FSO dual-hop transmission systems. IEEE Wireless Communications Letters. Jun 2013;**2**(3) 551–354

[47] Ferreira RM, Shahpari A, Reis JD, Teixeira AF. Coherent UDWDM-PON with dual-polarization transceivers in real-time. IEEE Photonics Technology Letters. Jun 2017;**29** (11):909–912

Chapter 3

Dielectric Resonator Nantennas for Optical Communication

Waleed Tariq Sethi, Hamsakutty Vettikalladi,
Habib Fathallah and Mohamed Himdi

Additional information is available at the end of the chapter

http://dx.doi.org/10.5772/intechopen.69064

Abstract

Dielectric resonator antennas (DRA) are ceramic based materials that are nonmetallic in nature. They offer high permitivity values (ε_r: 10-100). DRAs' have made their mark in various applications specially in the microwave and millimeter wave (MMW) spectrum, and are making encouraging progress in the THz band, because of their low conduction losses and higher radiation efficiencies compared to their metallic counterparts. With the advancements in nano fabrication, metallic antennas designed in the THz band have taken an interest. These antennas are termed as optical antennas or nantennas. Optical antennas work by receiving the incident electromagnetic wave or light and focusing it on a certain point or hot spot. Since most of the antennas are metallic based with Noble metals as radiators, the conducting losses are huge. One solution that we offer in this work is to integrate the nantennas with DRs. Two different DR based designs, one triangular and other hexagonal, are presented. Both the antennas operate in the optical C-band window (1550 nm). We design, perform numerical analysis, simulate, and optimize the proposed DR nantennas. We also consider array synthesis of the proposed nantennas in evaluating how much directive the nantennas are for use in nano network applications.

Keywords: dielectric resonator (DR), optical nanoantenna, hexagonal DR nantenna, triangular nantenna, optical communication C-window

1. Introduction

Over the last few years, with the introduction of various portable and wireless handheld devices, a surge in mobile and internet data traffic has been observed. This drastic increase is also effected by the way our society creates, shares and consumes information on a regular basis. These abundant data and information sharing also demand for increase in delivery time. High data rate communication with compact size of device is the new norm in technology. Researchers and industry are working at a fast pace to fulfil their consumers demands. It is estimated that wireless data rates are getting doubled every year and are quickly approaching the provided capacity of the wired communication systems [1–5]. Following this trend, systems working at higher data rates are needed. Although millimetre waves and 60 GHz radio [6–9] currently provide a solution and are being implemented in 5G applications, still researchers have to think ahead of time. The electromagnetic spectrum has a lot of bands to offer in terms of wide bandwidth. One section of the spectrum that has not been explored completely in the Terahertz (THz) band can offer communication in the Terabit-per-second (Tbps) domain. These (Tbps) links can be realised over the next 10 years. Terahertz band communication [10–14] is intended as a key wireless technology to satisfy this growing demand of bandwidth-hungry devices with requirements of higher data rates. Since THz band is in the early exploring phase, a lot of revisions and new standards have to be designed for the systems operating at these higher bands of the electromagnetic spectrum (ultraviolet band-infrared band). Special considerations have to be made in the design of transmission and receiving portion of the THz system [15].

Antenna is an important component of any wireless system. For the antenna to be designed at this higher end of the spectrum (THz band), special tools are needed from its realisation to characterisation. The antennas designed at this spectrum are termed as optical nanoantennas [16]. Optical nanoantennas work on the operating principle that the electromagnetic (EM) wave or the light wave received can be controlled or placed into localised energy spots pertaining to the design of the nanoantenna. This property of the optical nanoantennas have gained immense interest from the research point of view as it can be applied to various fields of applications, such as spectroscopy, sensing, photodetection, metasurfaces, medicine, photovoltaics and energy-harvesting applications[17–20]. With the advancements in nanotechnology fabrication, the nanoantennas designed at THz band can be realised. Nanotechnology is a fast-growing research area that marked use of machines that can fabricate nanocomponents. It is considered as an enabling technology for a set of applications in biomedical, environmental and military fields as shown in **Figure 1**. Being inherently simple and performing primitive operations only, nanomachines in isolation are not expected to manage advanced tasks. To enable more complex applications such as intra-body drug delivery or cooperative environment sensing, the exchange of information and commands between networking entities and/or external controller is required. The need for coordination and information sharing naturally leads towards the concept of nanonetworks. One promising way to enable networking capabilities is to use wireless communications between nanomachines [21] made possible with optical nanoantennas fabricated on these machines.

Optical nanoantennas present some similarities with their radio frequency (RF) counterparts, yet there still exists some major difference. The main challenge arises from the fabrication tolerances at nanoscales and from the drastic deviation of metals from perfect conductors to lossy plasmonic materials at optical frequencies [22]. This is usually described as the well-known dispersive plasmonic effects and results in a significant decrease of radiation efficiency caused by conduction losses. These losses can be theoretically explained via different models at

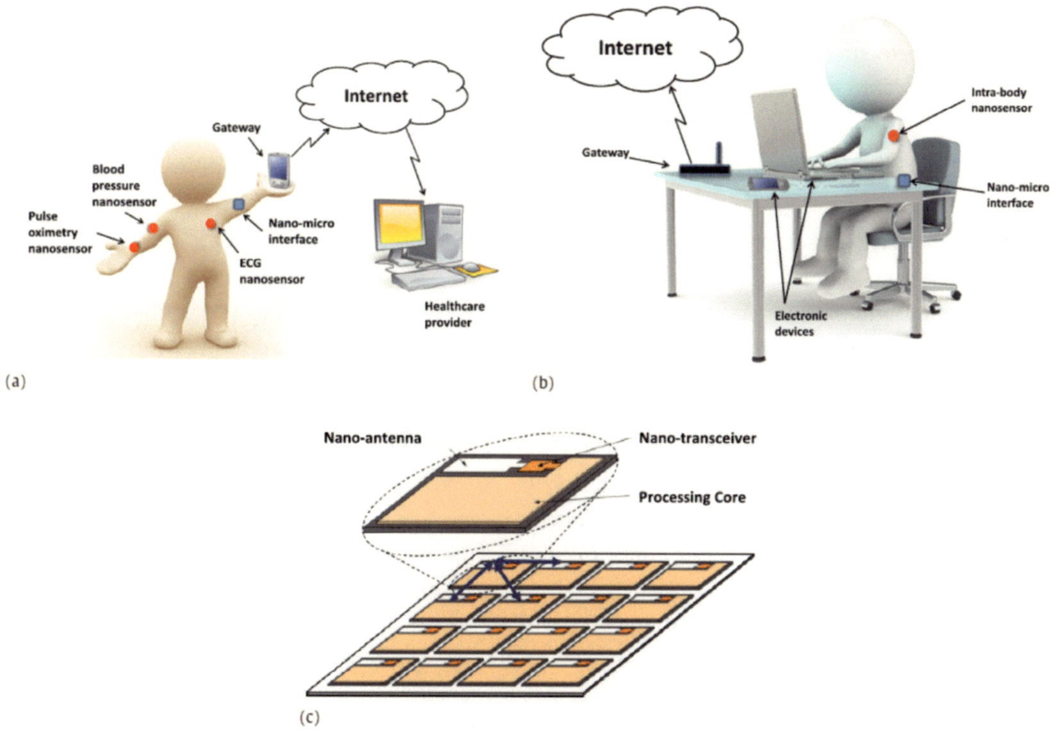

Figure 1. Applications for terahertz band (nanoscale antennas) [15].

specific frequencies with mostly known models such as Drude and Lorrentz model [23, 24]. One solution that is proposed in this chapter to cope with these metallic losses at high frequencies and to fully utilise the properties of metallic nanostructures is to design resonators or absorbers made with dielectric-based materials such as ceramics.

The idea of using antennas based on dielectric resonators (DR) was first proposed by Long et al. in 1983 [25], and since then the research into this novel idea is steadily increasing among the antenna researchers. The operation of DRAs as radiators exploit the 'radiation losses' of dielectric resonators made of moderate to high relative permittivity ($5 < \varepsilon_r < 100$) when excited in their lower order resonant modes in an open environment. Over the years, various aspects of DR antennas have been published and designs patented [26, 27]. Most notable publications are seen at the antennas operating as filters and circuits integrators at the microwave regime due to their compact size and high resistibility to losses. One striking feature of DR resonators is that they are immune to ohmic losses, which drastically appear at higher frequencies (bands above millimetre waves till far infrared), making them suitable candidates for optical nanoantennas. DRs are very efficient radiators compared to metallic antennas even at higher frequencies [28]. In this context, the work presented in this chapter focuses on integrating the resilient DR with metallic nanoelements for operating at THz band. The losses incurred due to appearing of plasmonics inside the metal elements are subsided with the integration of various DR-shaped resonating elements. The results presented in the later sections validate the selection of DR as a suitable candidate and replacement to traditional metallic resonators working at higher THz band.

2. State-of-the-art antennas

Optical nanoantennas are an attractive area for research in the field of optics and nanophotonics. With the advent of nanofabrication machines, the antennas designed at the lower RF and microwave millimetre wave (MMW) domains can now be scaled up to the THz domain. Thus, some of the properties and analysis obtained for antenna designs at the lower band can be applied to higher bands. To study the effects and analyse the current distributions on the optical nanoantennas, a new branch of physics emerged known as nanophotonics. Nanophotonics studies the transmission and reception of optical signals by submicron and even nanometre-sized objects. For nanooptics, it is important to efficiently detect and direct the transmitting signals for optical information between nanoelements. The sources and detectors of radiation in nanooptics are nanoelements themselves, their clusters and even individual molecules (atoms, ions). Nano objects functioning as antennas must exhibit high radiation efficiency and directivity [29, 30]. Most of the nanoantennas existing in the literature are based on plasmonic metallic structures (**Figure 2**). In Ref. [31], dipole nanoantennas exhibit the electric field localisation at certain spots, whereas the bowtie nanoantennas presented in Ref. [32] present broadband characteristics; and in Ref. [33], Yagi-Uda displays high directive nature which can be useful for nanonetwork communication among nanoscaled devices. Similarly in Refs. [34, 35], plasmonic nanoantennas provide enhanced and controllable light-matter interactions and strong coupling between far-field radiation and localised sources at the nanoscale. In Refs. [36, 37], magneto-plasmonic response of the nanoantennas is observed when ferromagnetic metals are driven not only by light but also by external magnetic fields. The authors observed that the magneto-plasmonic nanoantennas enhance the magneto-optical effects, which introduces additional degrees of freedom in the control of light at the nanoscale. However, regardless of various advantages of plasmonic nanoantennas associated with their small size and strong localisation of the electric field, such nanoantennas have large dissipative losses resulting in low radiation efficiency.

To alleviate this conduction loss phenomenon, we propose a combination of DR and plasmonic metallic-based optical nanoantennas. Literature review shows some of the existing designs that perform well when working with DRs. Since DR offers high dielectric constant and refractive index values of the materials, the losses are minimised and subsided when integrated or placed with metallic resonators. In rest of the sections, we detail the proposed DR-based nanoantenna designs with simulated results. The enhancement in the directivity of the nanoantenna is also discussed by implementing an array structure of 1×2 ETDRNA elements. Also, the tunability of the nanoantenna array is discussed. Finally, the chapter ends with the conclusion section discussing the presented results based on selected DR design geometries.

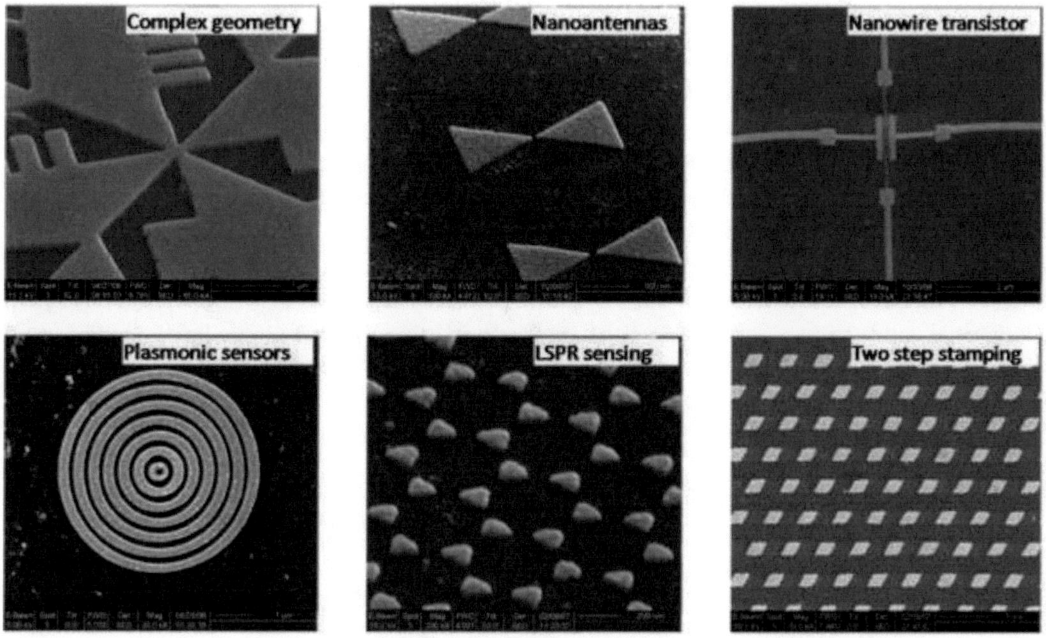

Figure 2. Different types of plasmonic nanoantennas: (a) complex connection, (b) bowtie, (c) dipole nanowires, (d) spiral sensor, (e) bowtie sensors and (f) rectangular patches [38].

Although we know that realization of any device is the ultimate proof of its operation, but at these very high THz frequencies it is difficult to have them fabricated as per limitations in fabrication resources. Secondly, our proposed designs are very small in dimensions (as will be discussed in later sections) to be realized currently with the exiting nano fabrication tools. Therefore we present the simulated designs using two different DR based resonators that we think will be viable candidates for designers and researchers who are in need of antenna designs having minimum losses and better radiation characteristics in the optical communication C-band at 1550 nm.

3. Equilateral triangular dielectric resonator nantenna

In this section, we present the simulated design and analysis of an equilateral triangular dielectric resonator nantenna (ETDRNA). The proposed nantenna is composed of a multilayer 'Ag-SiO_2-Ag' structure with noble metal silver (Ag) working as a feed transmission line. The dielectric triangular is made of silicon (Si) material and is excited via coupling mechanism from the feed line. The antenna yields a wide impedance bandwidth of 2.58% (192.3–197.3 THz) with a high directive radiation pattern of 8.6 dBi at 193.5 THz (1550 nm) with an end-fire radiation pattern.

3.1. Antenna geometry

Figure 3 shows the configuration (cross-sectional and front view) of the proposed equilateral triangular dielectric resonator nantenna (ETDRNA). The nanoantenna is designed to operate as a receiving antenna that can capture energy from the free space. The operating band of interest lies in the standard optical communication band at a wavelength of 1550 nm, which corresponds to central frequency of 193.5 THz. From the geometric configuration presented in **Figure 3**, the design follows a basic multilayer substrate approach. A silicon substrate (Si) having an oxide layer (O_2) is sandwiched between two conducting materials layers. The SiO_2 substrate has properties of thickness of h_1 = 0.150 μm, ε_r = 2.09 and loss tangent $tan\ \delta$ = 0 [39]. The partial conducting material below the substrate acts as a ground plane. Its dimensions and

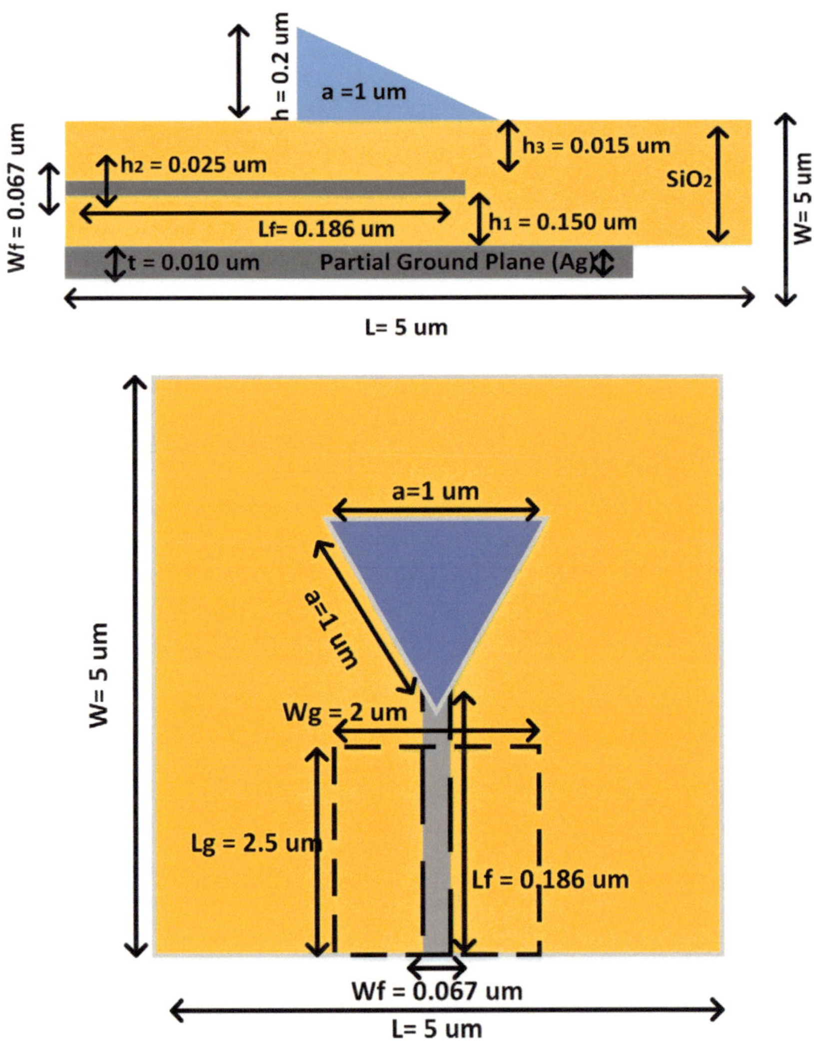

Figure 3. Antenna geometry of proposed nantenna design based on equilateral triangular DR (ETDRNA): (a) cross-sectional view; (b) front view.

thickness are $W_g \times L_g$ having a thickness of $t = 0.010$ µm. The nanoantenna is fed via a feed line placed on the top side of the substrate. It has geometric dimensions and thickness of $W_f \times L_f$; h_2 = 0.025 µm. The ground and the nanostrip are made up of noble metal silver (Ag). The dimensions of the substrate are taken as $W \times L = 5 \times 5$ µm^2. The resonator, made from (Si), placed on top of the feed line is made from dielectric material. It has the shape of an equilateral triangle with properties as $\varepsilon_r = 11.9$ and estimated loss tangent $tan\ \delta = 0.003$ at 100 THz [40]. The nanoantenna is excited and matched considering 50 Ω impedance source. In order to control the matching at the central frequency of 193.5 THz and to achieve a wide bandwidth with acceptable radiation patterns, the same (SiO_2) substrate material with thickness $h_3 = 0.015$ µm has been introduced between the equilateral triangle and the nanostrip. The dimensions of the equilateral triangular dielectric are calculated from Eq. (1) [41, 42].

$$f_{mnl} = \frac{c}{2\sqrt{\varepsilon_r}}\left[\sqrt{\left(\frac{4}{3a}\right)^2 + \left(\frac{p}{h}\right)^2}\right]^{1/2} \tag{1}$$

where 'a' is the side length of the equilateral triangular DRA, ε_r is the dielectric constant of the DRA, 'h' is two times the height of the triangular DRA to account for the image effect of the ground plane and $p = 1$ for the fundamental mode [42]. For a low-profile triangular DRA, we have $a >> h$, and therefore Eq. (2) demonstrates that the frequency is predominantly determined by the height of the DRA:

$$fr = \frac{c}{4h\sqrt{\varepsilon_r}} \tag{2}$$

where h and ε_r are the height and dielectric constant of triangular DRA.

Metals working in the optical regime are faced with another loss. This loss appears as negative permittivity, therefore complex permittivity ε_{Ag} of the metals, in our case silver (Ag), is calculated from Eq. (3) explained by the Drude model which is based upon kinetic theory of electron gas in solids [39]:

$$\varepsilon_{Ag} = \varepsilon_o\left\{\varepsilon_\propto - \frac{f_{p^2}}{[f(f + i\gamma)]}\right\} = -129.17 + j3.28. \tag{3}$$

where ε_o = 8.85 × 10^{-12} [F/m], $\varepsilon_\propto$ = 5, plasmonic frequency f_p = 1.41e^{16} rad/s, f = central frequency and collision frequency γ= 2.98e^{13}. The plasmonic frequency, which appears after the photon and free electron gas collision, defines the collective motion of the electrons and can be expressed as follows:

$$f_p = \sqrt{ne^2/\varepsilon_0 m} \tag{4}$$

where n is electron concentration, e is the free electron charge (1.6 × 10^{-19} C), ε_0 is the free space (vacuum) permittivity (8.854 × 10^{-12} F/m) and m is the electron effective mass. From

Eq. (4), the behaviour of arriving EM wave to the metal can be deduced. For $f < f_p$ corresponds to and exponential decay field, EM wave will be reflected back and will not propagate through the metal. On the other hand, if $f > f_p$, the EM wave will behave as a travelling wave and will pass through the metals. Similarly, the collision or damping factor describes the losses within the metal and can be expressed as:

$$\gamma = \frac{e}{\mu m} \tag{5}$$

where μ is mobility of free carriers. The proposed model has taken into account the conductive and dielectric losses and has been simulated in commercially available EM simulator CST MWS 2014 based on FIT numerical technique using optical template.

3.2. Design simulation and optimisation

To get an understanding on the working principle of the proposed ETDRNA, various parameters of the nanoantenna were extensively optimised. In order to study the effects of the antenna performance in terms of bandwidth and directivity, the following parameters were observed and analysed.

a. **Nanostrip feed:** The nanostrip feed line placed on top of the SiO_2 substrate was optimised in terms of its length and width. The traditional empirical formulas [43] were used as a starting point for the nanostrip design. The nanostrip acts like a coupling resonator that excites the triangular dielectric place on an upper SiO_2 substrate with height h_3.Traditionally at RF frequencies, the length of the transmission lines are characterised to the wavelengths (λ) of incoming and outgoing radiations. However, working at the optical frequencies, the incident waves reflect less and penetrate more in to the substrate passing through the metal atoms. This phenomenon is known as plasmonic affect, and it gives rise to free plasmonic gaseous atoms. To deal with this new phenomenon at optical frequencies, we use shorter effective wavelength (λ_{eff}) compared to traditional wavelengths (λ), which depends on material characteristics given by Eq. (6) for length of a transmission line [44]:

$$\frac{m\lambda_{eff}}{2} = L(\lambda_o) \tag{6}$$

where Eq. (6) shows the relationship between the free space wavelength (λ_o) and the effective wavelength (λ_{eff}) and the order of resonance (m). Here, effective wavelength is given by:

$$\lambda_{eff} = \frac{\lambda_o}{n_{eff}} \tag{7}$$

Typical values of n_{eff} has been measured to be in the range of 1.5–3 [45]. In our case for the silver nanostrip feed line, we use $n_{eff} = 2.8$ which resulted the minimum resonating length of to be 0.27 μm. The length L_f of the nanostrip was optimised from 0.1 to 0.27 μm with the best optimised value producing required resonance at 193.5 THz was at $L_f = 0.186$ μm as shown in **Figure 4a**. The effect of the width 'W_f' of the nanostrip was also examined by extensive parametric studies. Initial values were taken from the empirical formulas [43] and optimisation was done from 0.02 to 0.28 μm. **Figure 4b** shows the best optimised value achieved at resonance of −22 dB with W_f = 0.067 μm.

b. **Partial ground plane:** The ground plane plays an important role in controlling the bandwidth and radiation characteristics of any designed antenna. At the nanoscale geometry, we simulated and observed its parameters effect on our nanoantennas resonance behaviour. We started off initially with a finite ground plane that achieved a good radiation pattern with an acceptable bandwidth. The ground plane was then optimised and a partial section of it was used with optimised dimensions $L_g \times W_g$ = 0.5 μm × 2 μm. **Figure 5a** and **b** shows the effects of varying the ground plane in terms of its length and width. The optimised results produce a wide impedance bandwidth of 2.5% (192.3–197.3 THz) at a centre frequency of 193.5 THz. This makes our proposed nanoantenna covers all the standard optical transmission widow (C-band), with a directivity of 8.6 dB.

c. **Height of triangular DR:** Since the height of the triangular DR predominately determines the resonance frequency as according to Eq. (2), the height h of the DR was optimised from 0.1 to 0.5 μm. **Figure 6** shows the best optimised value of h = 0.3 μm having a resonance at −23 dB.

d. **Rotation of triangular DR:** In order to study the effects of bandwidth, frequency shift and directivity of the nanoantenna design, the proposed silicon-based triangular DR was rotated on its axis. The rotation was from 0° to 360° with an angular spacing of 40°. **Figure 7** shows the angular rotation of the triangular DR. The tip of the triangle was initially aligned at 0° shown in green colour. The DR was then rotated along the counter-clockwise direction with varying angles. It was observed that with the rotation of the DR, the bandwidth remained the same at 2.5% but the resonant frequency shifted to other bands (200–205 THz) in the frequency range from (180–220 THz) as shown in **Figure 8a**. Since the triangle is an equilateral one, the angular rotation produces the same shifts at

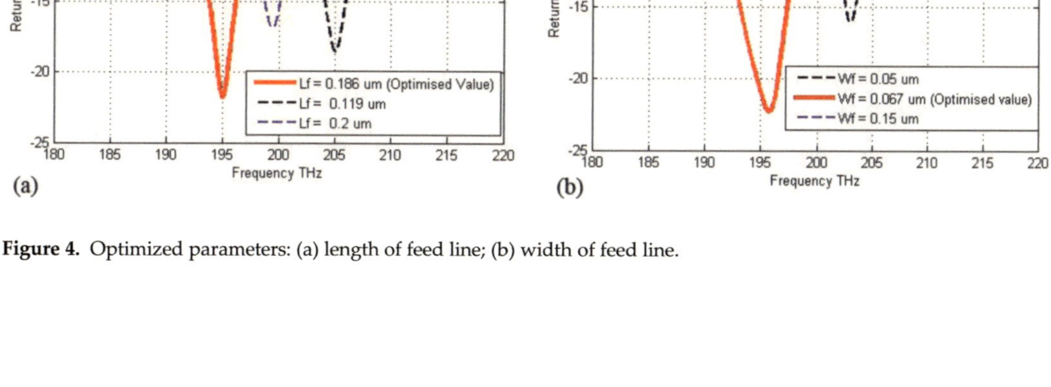

Figure 4. Optimized parameters: (a) length of feed line; (b) width of feed line.

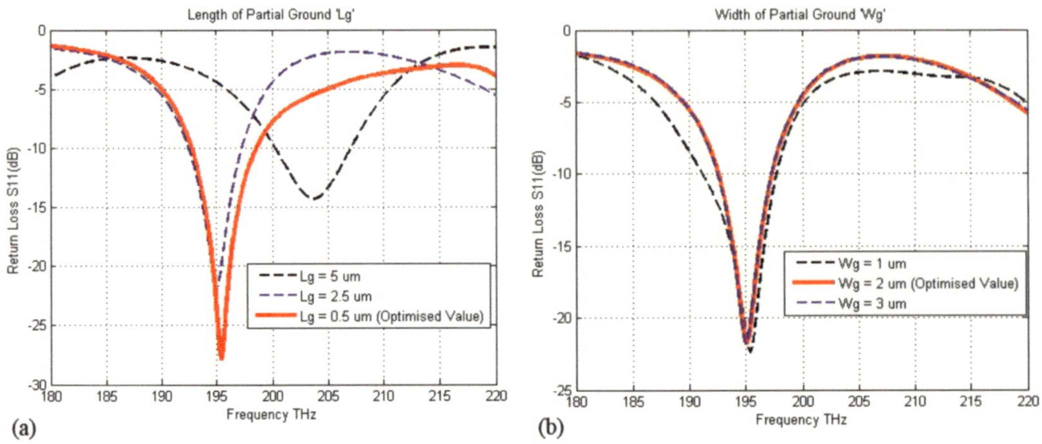

Figure 5. Optimized parameters: (a) length of ground plane; (b) width of ground plane.

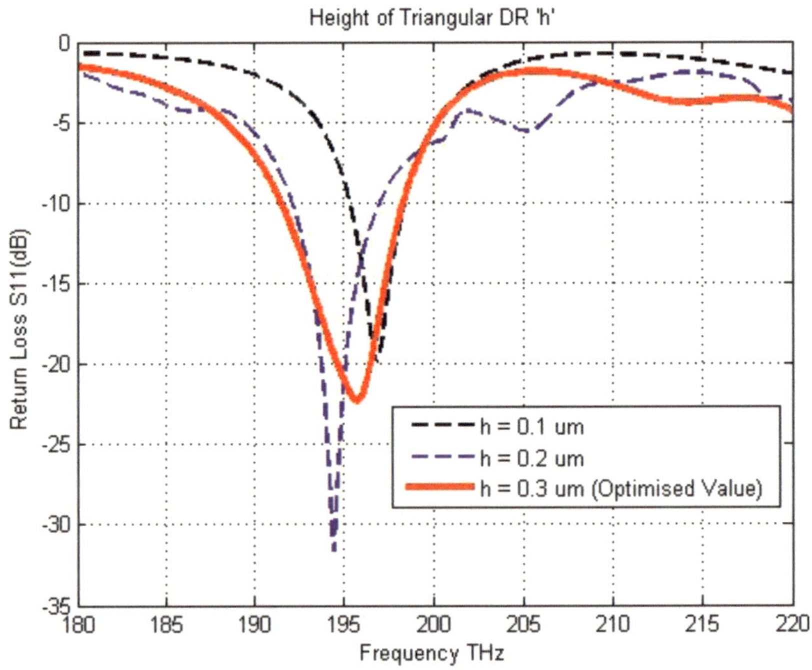

Figure 6. Optimized parameters: height of triangular DR.

other angles, that is, the shift will be the same at (0 = 120 = 240 = 360°) as shown in **Figure 8b**. The directivity was also affected with the rotation of the triangle as shown in **Figure 8b**. It is clear that the effect of the rotation of the triangular DR lowers the directivity to nearly 3 dBi.

http://dx.doi.org/10.5772/intechopen.69064

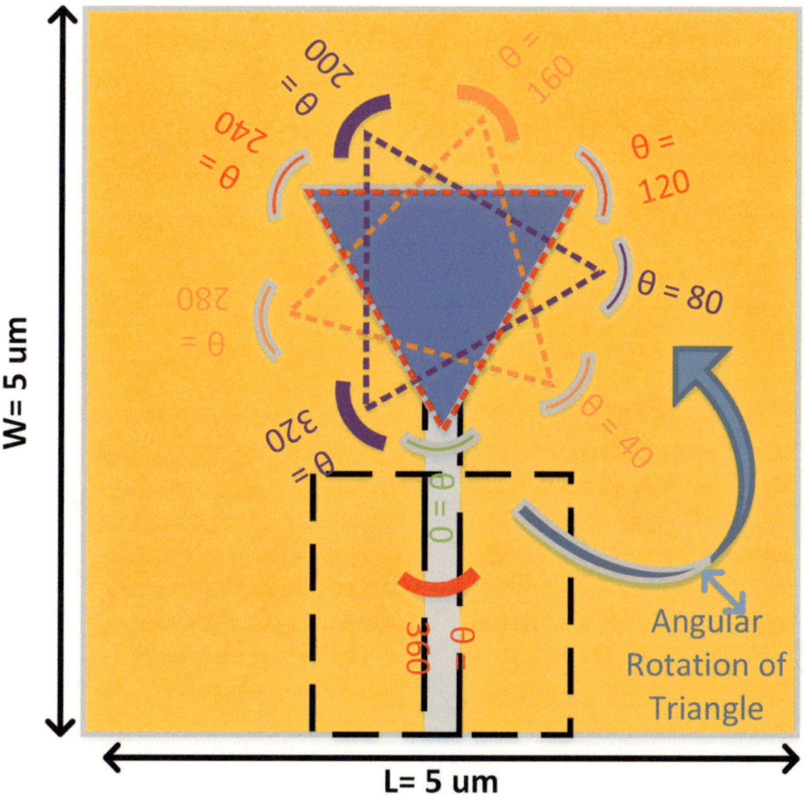

Figure 7. Angular rotation of triangular DR.

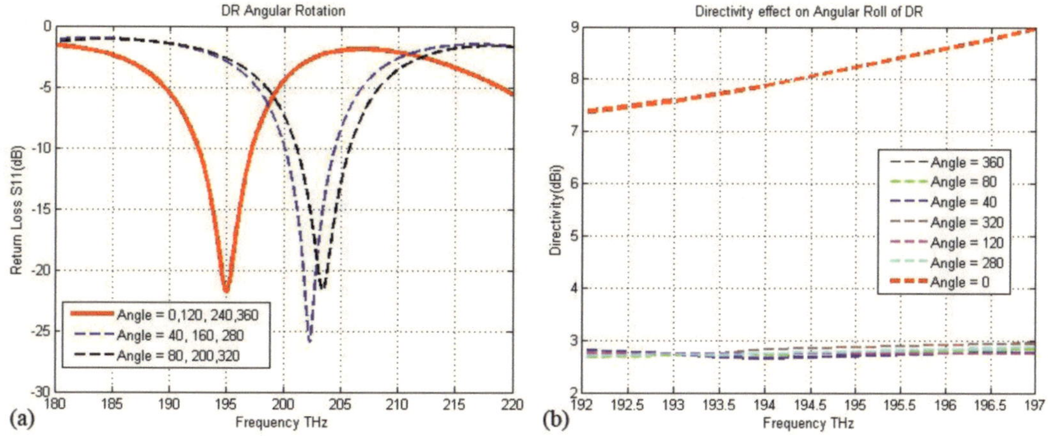

Figure 8. Optimized parameters: (a) effect of angular rotation of resonant frequency; (b) effect of angular rotation on directivity.

Parameters	Value (μm)
Feed length L_f	0.186
Feed width W_f	0.067
Ground length L_g	2.5
Ground width W_g	2
Height of triangular DR h	0.2
Area of triangular side a	1
Rotation angle θ	0^0

Table 1. Optimised parameters of ETDRN.

After extensive optimisation of stated parameters above and analysing the results achieved from these optimisations, the best geometric parameters that achieve an impedance bandwidth of 2.5% (192.3–197.3 THz) and a directivity of 8.6 dB are listed in **Table 1**. It is also observed that the simple ETDRNA structure can act as a tunable resonator when rotated around its axis resulting in usage of applications that work in the wavelengths in the range of (1463–1500 nm). The proposed design, if facility exists, can be fabricated via the known techniques involved in nanofabrication technology, that is, e-beam lithography, photolithography and chemical vapour deposition. In our case, the fabrication will follow a bottom-up approach where the SiO_2 substrate will have silver deposited on its surface.

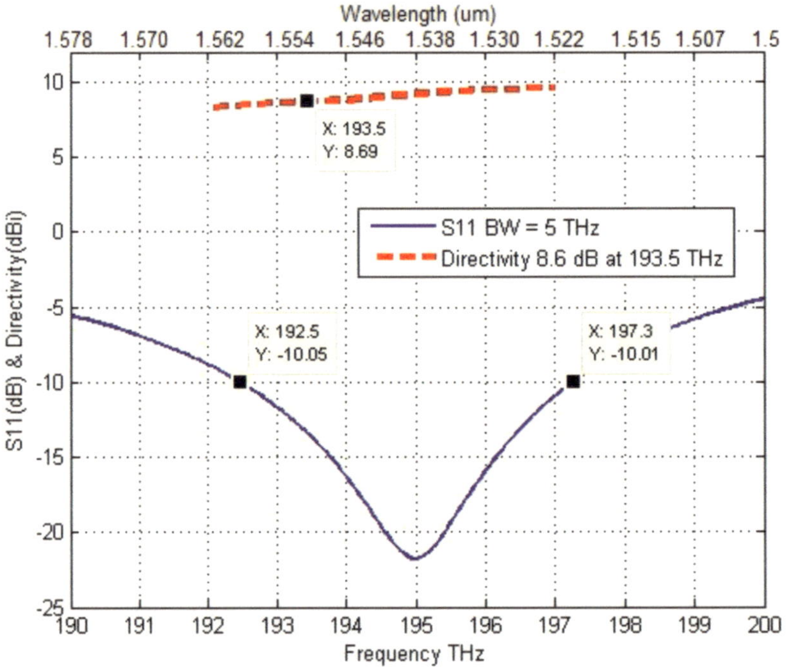

Figure 9. Simulated return loss and directivity of ETDRNA.

3.3. Results and discussion

In this section, we present the simulation results. **Figure 9** shows the return loss (S_{11}) and directivity of the proposed ETDRNA. The three-dimensional (3D) radiation patterns of the nanoantenna at 192, 193.5 and 197 THz are shown in **Figure 10(a–c)**. The maximum dip of −22 dB is achieved from the resonance of the nanoantenna at the central frequency of 1936.5 THz. The nanoantenna covers some part of the S-band while most part is covered for the C-band optical communication window. 3D radiation patterns provide the proof of the ETDRNA radiating in end-fire pattern. At present, the nanofabrication technology is limited and the proposed design is a theoretical one, yet we believe that our contribution in the fast-growing field of nantennas, with the proposed ETDRNA design, will prove itself to be a promising candidate for next-generation energy harvesting and green sustainable solution applications based on nanotechnology designs.

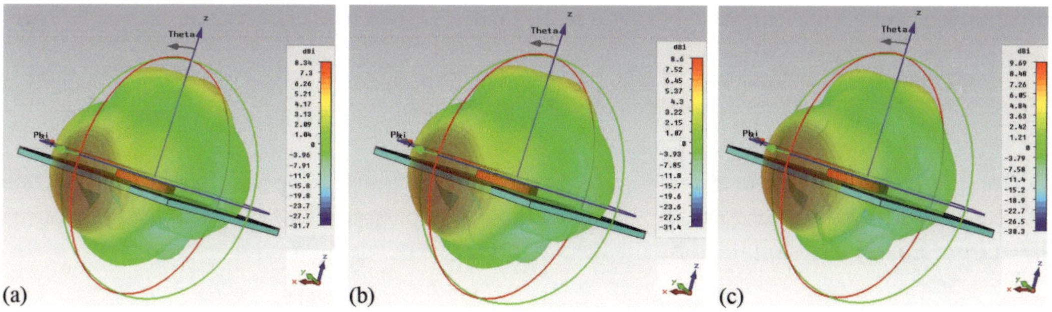

Figure 10. 3D end-fire radiation pattern at: (a) 192 THz, (b) 193.5 THz, (c) 197 THz.

4. Hexagonal dielectric resonator nantenna

In this section, we present another nantenna design based on dielectric resonator material. The shape of this DR is in hexagonal form with the material chosen as silicon (Si). The proposed design works as a loading element. The structure is again in multilayer form having (SiO_2) sandwiched between two silver (Ag) sheets. The radiating element is an equal-sided hexagonal-shaped (Si) dielectric loaded material. The whole nantenna structure is excited via a nanostrip transmission line made from a noble silver metal (Ag) whose conductive properties are calculated using the Drude model. The antenna achieves an impedance bandwidth of 3.7% (190.9–198.1 THz) with a directivity of 8.6 dBi at the frequency of interest. The obtained results make the proposed nantenna a possible solution for future nanophotonics and nanoscale communication devices.

4.1. Antenna geometry

In this section, we present another dielectric resonator (DR) design that takes the shape of a hexagon. The proposed nanoantenna works utilised the loading properties of ceramic dielectric silicon and is termed as hexagonal dielectric loaded nantenna (HDLN). It is also designed to operate at the central frequency of 193.5 THz which corresponds to a wavelength of 1550 nm. The cross-sectional and front view of the proposed HDLN is in **Figure 11(a)** and **(b)**. The design is based on a multilayer structure with (SiO_2) substrate sandwiched between two noble metals each made from silver (Ag). The properties of substrate are: thickness of $h_1 = 0.150$ μm, $\varepsilon_r = 2.1$ and loss tangent $tan\ \delta = 0.003$ at $f = 100$ THz [39]. The ground layer, made from silver, below the substrate has partial form with properties as thickness of $t = 0.010$ μm and dimensions $L_g \times W_g = 1.95 \times 2$ μm. The feeding line is a nano silver strip on top of the substrate with parameters: thickness $h_2 = 0.025$ μm, $W_f = 0.067$ μm and $L_f = 0.186$ μm. The substrate dimensions are taken as $W_s \times L_s = 5 \times 5$ μm^2. The hexagonal dielectric is made of (Si), with $\varepsilon_r = 11.9$ and estimated loss tangent, $tan\ \delta = 0.0025$. To achieve a further increase in the bandwidth with minimum resonance losses, a small substrate with thickness $h_3 = 0.015$ μm made from (SiO_2) has been introduced between the hexagon and the nanostrip. The dimensions of hexagonal dielectric are calculated from Eq. (8) [43] by inscribing the hexagon inside a circle and equating the areas of both designs, thus giving an optimised equal side lengths of hexagon as $s = 1$ μm and thickness ($\lambda_g/4 < h < \lambda_g/2$) $h = 0.377$μm;

$$\pi a_e^2 = \frac{3\sqrt{3}}{2} s^2 \tag{8}$$

where a_e = area of the circle and s = side of the hexagon. Since at optical frequencies, metals appear with a negative permittivity; therefore, complex permittivity 'ε_{Ag}' of silver (Ag) calculated from Eq. (9) was explained by the Drude model [39]:

$$\varepsilon_{Ag} = \varepsilon_o \left\{ \varepsilon_\propto - \frac{f_{p^2}}{[f(f + i\gamma)]} \right\} = -128 + j3.28 \tag{9}$$

where $\varepsilon_o = 8.85 \times 10^{-12}$ [F/m], $\varepsilon_\propto = 5$, plasmonic frequency $f_p = 2175$ THz, f = central frequency and collision frequency $\gamma = 4.35$ THz. **Figure 11(b)** illustrates the antenna operating in the transmitting (Tx) mode by means of propagation vector orientation (k). The magnetic and electric field distributions of the hexagonal dielectric and nanostrip waveguide, along with the wave propagation in the y-axis, are also shown. Optical nantennas can be excited with a few known techniques ; (1) coupling of light using the so called nanotapers [46–47] since nano antennas cannot handle much power because of their small footprints, this makes them ideal candidates for being excited by micro lasers such as micro disks and photonic crystal lasers and (2) by reducing the reflection induced power loss by using slot dielectric waveguides [48].

4.2. Design simulation and optimisation

In this section, we make use of the optimisation techniques available to us from the simulator. We investigate the performance of each parameter involved in the design geometry of the proposed nanoantenna as shown in **Figure 11**. In order to study the impact on the antenna performance in terms of bandwidth, the following parameters have been studied and analysed.

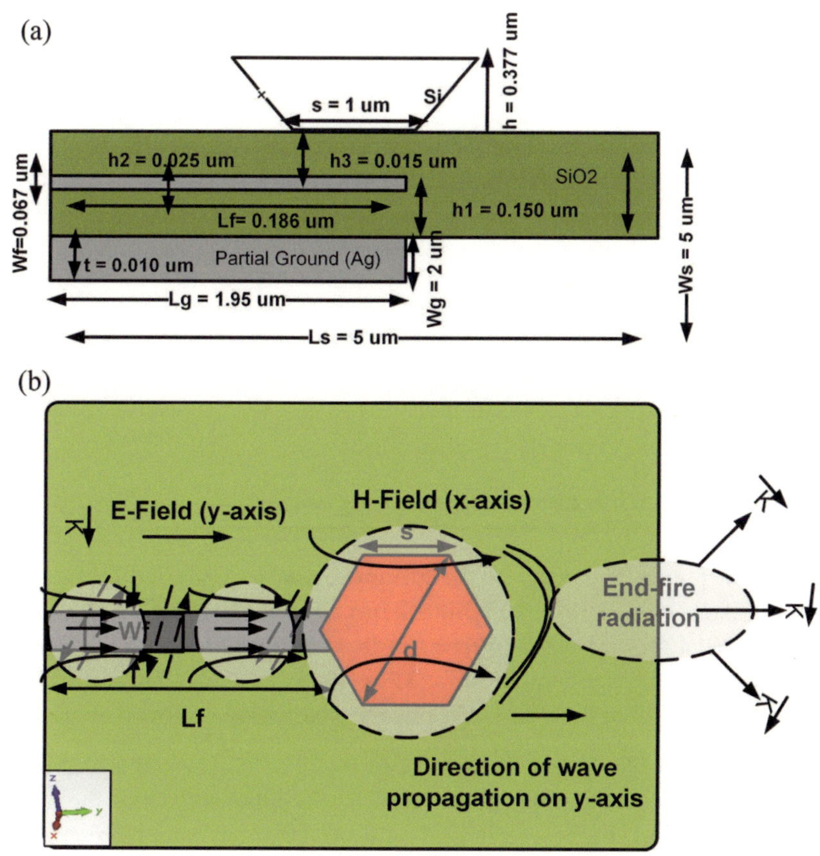

Figure 11. Geometry of hexagonal dielectric loaded nantenna (HDLN): (a) cross-sectional view; (b) front view with field vectors.

a. **Nanostrip feed:** Properties of conducting materials change when working at optical frequencies. The silver nanostrip line used to feed the nanoantenna was analysed in terms of Drude model. The nanostrip acts like a coupling resonator that excites the hexagonal dielectric, placed on an upper SiO_2 substrate with height h_3. Traditionally at RF frequencies, the length of the transmission lines is characterised to the wavelengths (λ) of incoming and outgoing radiations. However, working at the optical frequencies most of the incident light is transparent through the metals. This gives rise to plasmonic-free electrons, thus the feed line is analysed considering shorter effective wavelength (λ_{eff}), which depends on the material properties [44], refractive indexes and Eqs. (6) and (7). In our simulations, the optimised dimensions of the feed line produced values of length L_f to be 0.27 μm. The length L_f of the nanostrip stub was optimised from 0.1 to 0.27 μm with the best optimised value producing required resonance at 193.5 THz was at L_f = 0.186 μm as shown in **Figure 12(a)**. Similarly, the width 'W_f' of the nanostrip was also examined and optimisation was done from 0.02 to 0.28 μm. **Figure 12(b)** shows the best optimised value achieved at resonance of −22 dB with W_f = 0.067 μm.

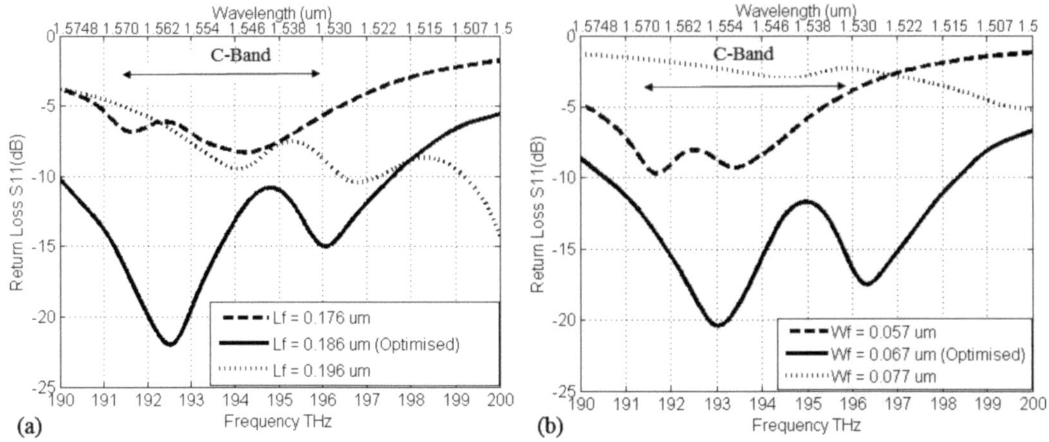

Figure 12. Optimized parameters: (a) length of feed line; (b) width of feed line.

b. **Partial ground plane:** The effect of the ground plane was studied and its optimisation gave dimewnsions of $L_g \times W_g = 1.95\ \mu m \times 2\ \mu m$. **Figure 13(a)** and **(b)** shows the effects of varying the partial ground plane in terms of its length and width. The optimised results produce a wide impedance bandwidth of 3.7% (190.9–198.1 THz) at a centre frequency of 193.5 THz, covering all of the standard optical transmission widow (C-band).

c. **Height of hexagonal DR:** The wide impedance bandwidth achieved is also affected by the height of the hexagonal DR. The height h of the DR was optimised within the range ($\lambda_g/4 < h < \lambda_g/2$) [3]. **Figure 14** shows the best optimised value of $h = 0.37\ \mu m$ having a resonance at -23 dB.

4.3. Results and discussion

To compare the properties and results of our proposed HDLN nanoantenna, we first simulated a hexagonal dielectric resonator antenna at the lower frequency band [41]. Observations were

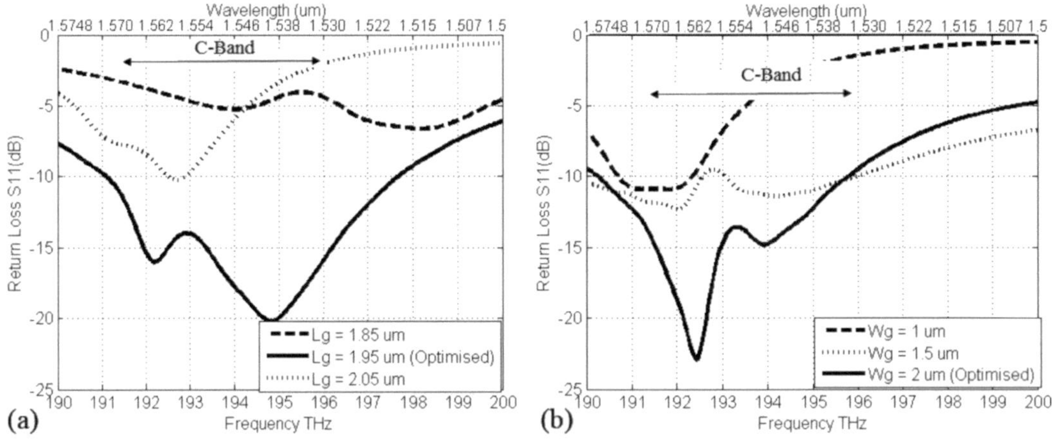

Figure 13. Optimized parameters: (a) length of ground plane; (b) width of ground plane.

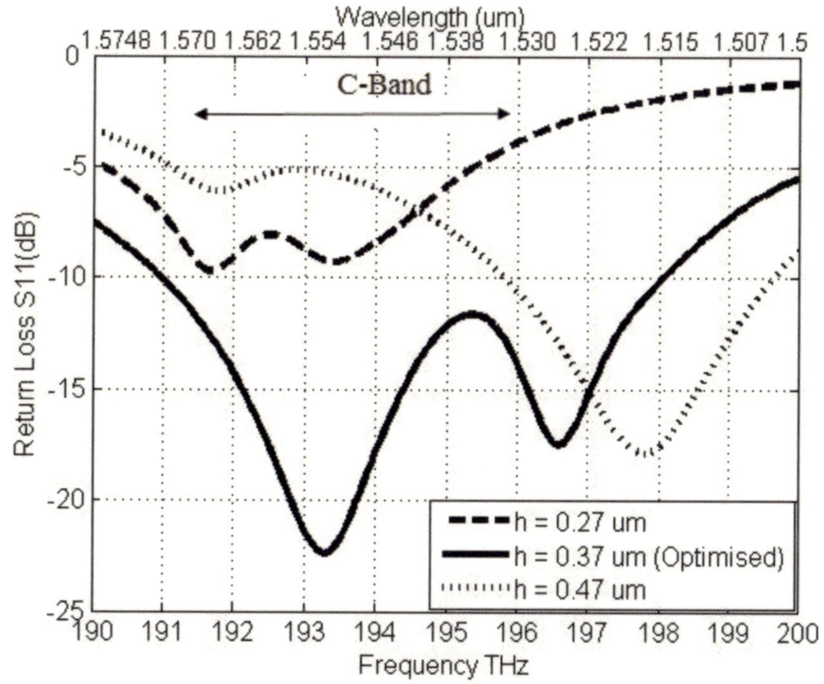

Figure 14. Optimised parameters: height of hexagonal DR.

made in terms of plane-wave propagation in the transmission lines to the radiating structures of the two antennas with results shown in **Figure 15(a)** and **(b)**, respectively. From **Figure 15 (a)**, it can be observed that the E-field propagation or the power propagation in the transmission line is following the fringing effects in order to radiate the hexagonal structure operating in the microwave domain. Whereas the proposed nantenna structure depicted in **Figure 15(b)** shows the E-field propagation in the nanotransmission line follows a travelling wave effect. It is also observed that the hexagonal DR elements for both the cases exhibit different properties. At the microwave domain, the hexagonal DR as shown in **Figure 15(a)** works as a resonator,

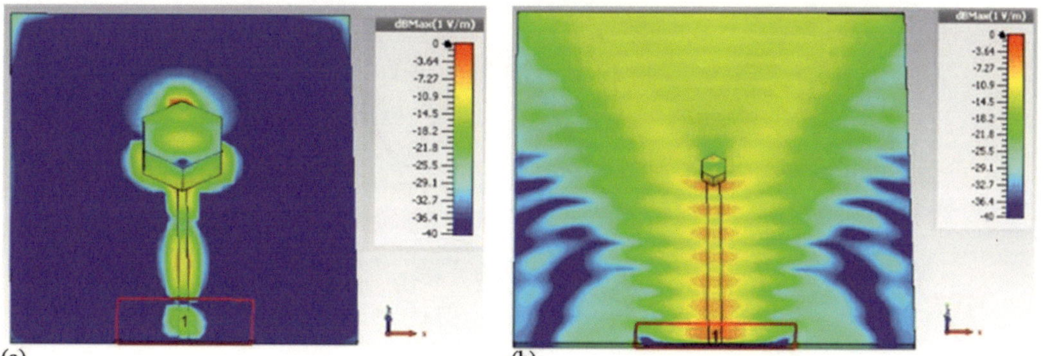

(a) (b)

Figure 15. E-field propagation: (a) fringing effects at lower frequency; (b) travelling wave effect at THz spectrum.

whereas the DR at the nanoscale structure shown in **Figure 15(b)** exhibits loading properties which benefit the nantenna to operate as a lens and thus achieve more directivity.

The return loss (S_{11}) and directivity of the proposed HDLN with respective wavelength and directivity axis is shown in **Figure 16**. After extensive optimisation, the nantenna achieves an impedance bandwidth of 3.7% (190.9–198.1 THz) with a directivity of 8.6 dBi, making it useful for nanoscale fabrication due to its robustness against fabrication tolerances.

Typically, the modes of hexagonal DR [49] are derived from the cylindrical dielectric resonator, which has three distinct types: TE (TE to z), TM (TM to z) and hybrid modes. The TE and TM modes are asymmetrical and have no azimuthal variation. On the other hand, fields produced by hybrid modes are azimuthally dependent. Hybrid mode is further divided into two sub-groups of HE and EH. The modes generated by the hexagonal dielectric nantenna are represented in terms of magnitude of electric field distribution on its surface as shown in **Figure 17**, at the centre frequency of 193.5 THz. The mode analysis was done via EM simulator CST MWS.

From the infinite modes available [50], in our simulation as shown in **Figure 17**, we observed the nano hexagonal dielectric antenna producing $HE_{20\delta}$ mode between the achieved wide impedance band. The subscript in the modes represents the variation of fields along azimuthal, radial and z-direction of the cylindrical axis. It is observed from the figure that the magnitude of electric field variation is produced on the azimuthal direction with no variation in the radial direction, thus giving a mode excited at $HE_{20\delta}$. Also, the intensity is highest at the azimuthal

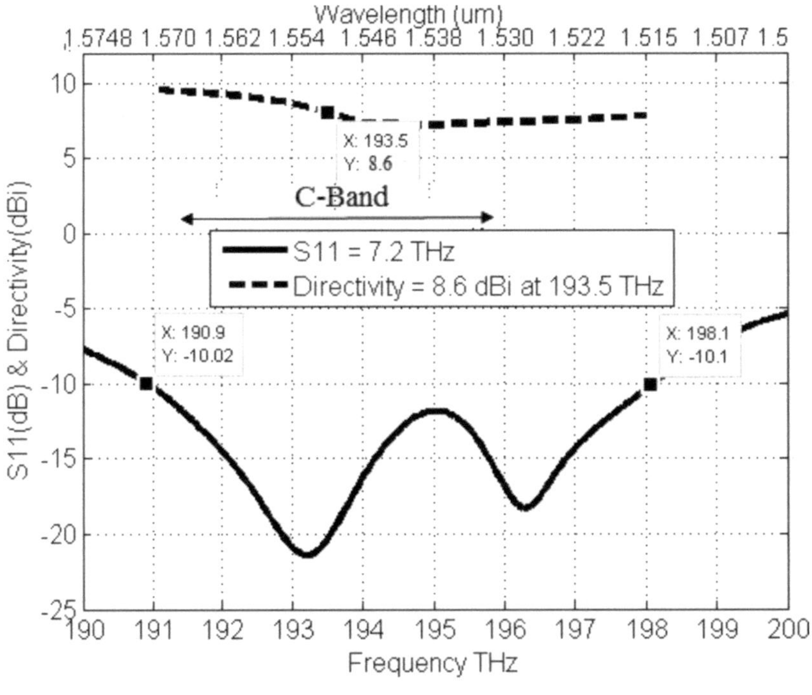

Figure 16. Simulated return loss and directivity of proposed HDLN.

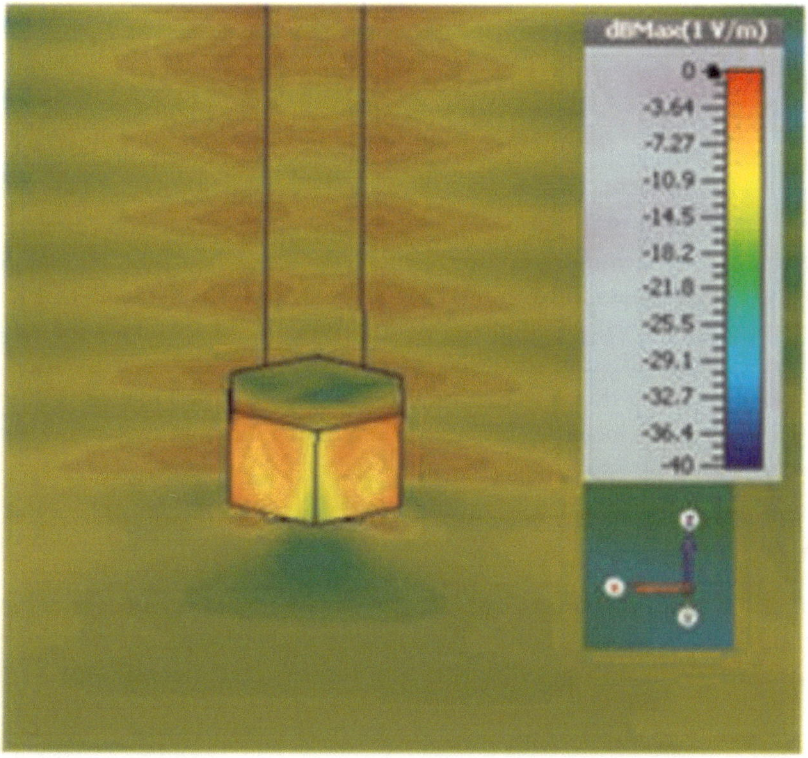

Figure 17. E-field distribution as shown via the magnitude at 193.5 THz with $HE_{20\delta}$ mode.

plane resulting in a radiation pattern towards the end-fire direction. The 3D radiation patterns of the nanoantenna at 191, 193.5 and 198 THz are shown in **Figure 18(a–c)**. The directivity of the antenna at the centre frequency is 8.6 dBi.

Keeping in mind the state-of-the-art nantenna designs and limited availability of nanofabrication equipment and facilities worldwide, we believe our proposed theoretical HDLN design will prove itself to be a promising communication device for applications based on nanotechnology.

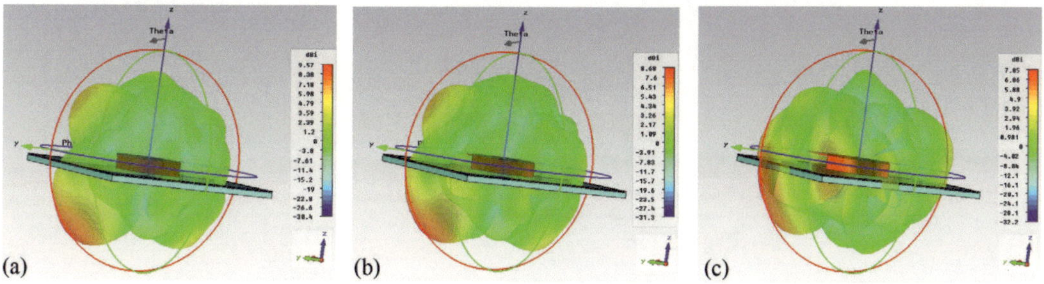

Figure 18. 3D radiation pattern: (a) 191 THz, (b) 193.5 THz and (c) 198 THz.

5. Array synthesis

In this section, we present the array synthesis done on one of the two proposed nantenna designs based on DR element. The ETDRNA was opted for array optimisation. The ETDRNA was fed via a 1 × 2 corporate feed network working at optical C-band (1.55 μm). Numerical results prove that the proposed nantenna exhibits a directivity of 9.57 dB with an impedance bandwidth of 2.58% (189–194 THz) covering the standard optical C-band transmission window. Furthermore, by selecting the appropriate orientation of the triangular dielectric resonators, the proposed nantenna structure can be tuned to operate at the higher or lower optical bands offering a threshold value of directivity and bandwidth Δ*f*. By tuning the nantenna, they achieve an increase in bandwidth of 4.96% (185.1–194.7 THz) and directivity also improves to 9.7 dB. The wideband and directive properties make the proposed nantenna attractive for a wide range of applications including broadband nanophotonics, optical sensing, optical imaging and energy-harvesting applications.

5.1. Antenna array geometry

An array of 1 × 2 configuration for the ETDRNA is presented. The front and side view is shown in **Figure 19**. The geometry of the proposed nanoantenna utilised the same parameters as the proposed ETRDN in Section 3. The gap between the elements to counter mutual

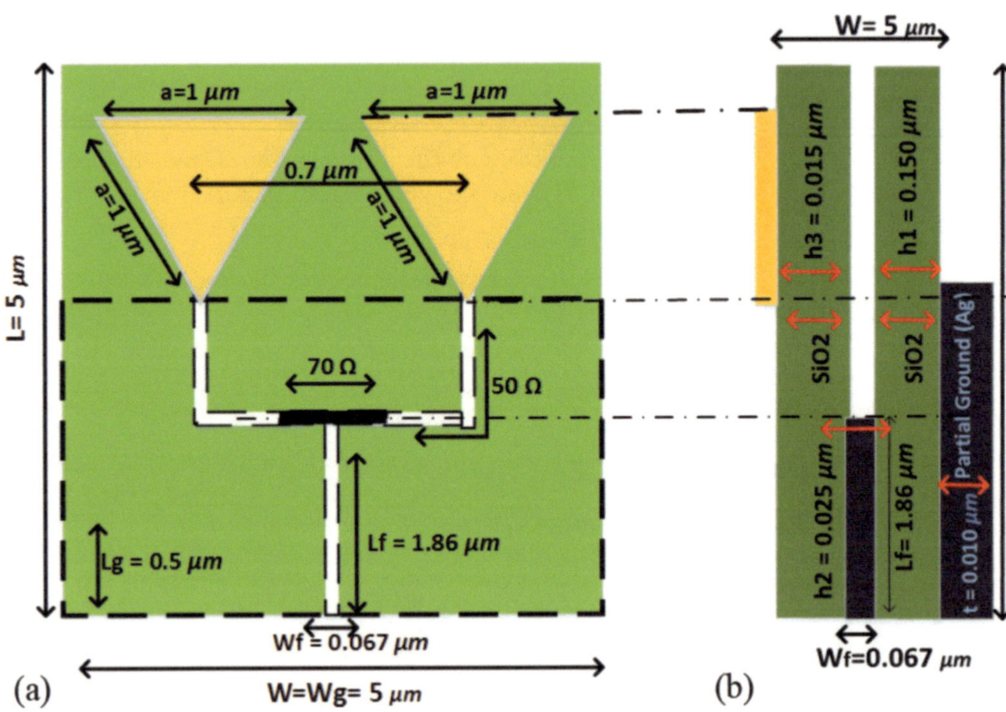

Figure 19. Geometry of proposed 1 × 2 array: (a) front view; (b) side view.

coupling is around ($\lambda_{eff}/2$) at the central frequency of 193.5 THz. The dimensions of the single element equilateral triangle dielectric based on frequency dependence can be calculated from Eqs. (1) and (2) previously presented in Section 3.

In order to increase the directivity of the proposed nantenna, arrays with a corporate feed network has been utilised for best power transfer from the source to the radiating ETDRNA. **Figure 19(b)** shows the corporate feed network along with appropriate feed line (50 and 70 Ω) markings. The optimised width of the 70 Ω feed line is 0.045 μm, whereas the length is 0.76 μm. For the 50 Ω feed line, the width and length are W_f and L_f, respectively. The shaded region in black shows the partial ground and feed lines to be on the backside of the substrate. The optimum distance between the triangular plasmonic resonators achieve minimal mutual coupling at 0.7 μm centre to centre. The properties of the noble metals are explained as per Drude Model used throughout the sections for the proposed nantenna designs.

5.2. Results and discussion

In this section, we investigate the proposed nanoantenna arrays results in terms of two important features: (1) directivity improvement and (2) tunability. Directivity is very important because it measures the power density the antenna radiates in the direction of its strongest emission versus the power density radiated by an ideal isotropic radiator (which emits uniformly in all directions) radiating the same total power, and on the other hand tunability is an attractive feature as it makes antenna operational in various frequency bands simultaneously as well as resonating at its centre frequency. In fact, during our performance analysis, we learned the importance of the angle of rotation between the triangular structure and the feed line direction. In the following paragraph and the next section, we first address the performance study when this angle is zero and then investigate the importance of this when different angles are introduced.

The simulated return loss (S_{11}) and directivity of the proposed 1 × 2 nanoantenna array are presented in **Figure 20**. The proposed nantenna achieves an impedance bandwidth of 2.58% with $S_{11} < -10$ dB (189–194 THz) at a centre frequency of 193.5 THz (1550 nm) with a dip at −15 dB. The antenna covers all the portion of the standard C-band transmission window in optical domain and can be used for relevant optical applications in nanonetworks and high-speed optical data transfer. The 3D radiation plot of the nantenna resonating at 193.5 THz is shown in **Figure 21**. The directivity of the nantenna is 9.57 dBi, 0.97 dB improvement compared to the single element ETDRNA discussed in Section 3. CST MWS studio has been used to acquire the optimised results with verification done by another EM simulator HFSS. Examining the plot in **Figure 21** reveals the E-field component having main lobe direction at 45°, side lobe levels at −4.1 dB and beam width of 20.1°. Similarly, the H-field component in **Figure 21** has main lobe direction of 180°, side lobe levels of −1.5 dB and beam width of 127.2°. Although the nantenna achieved a directive nature, the losses associated with high side lobe levels are to be expected due to substrate selection and working at higher THz frequencies.

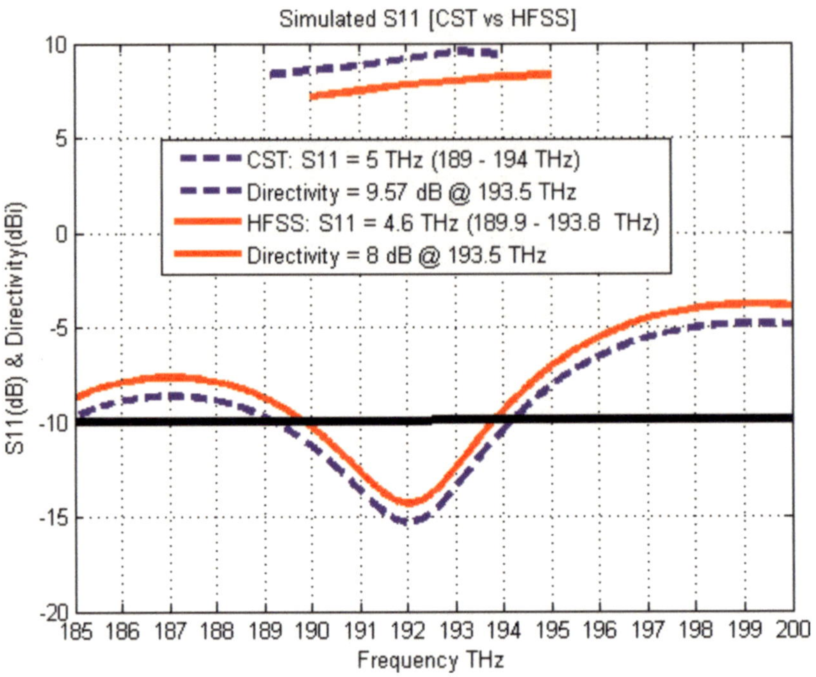

Figure 20. Return loss and directivity of 1×2 ETDRNA array.

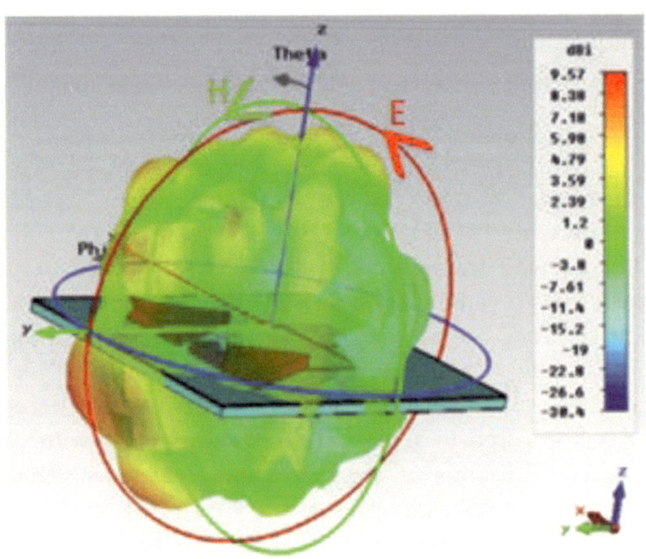

Figure 21. 3D radiation pattern at 193.5 THz.

5.3. Tunability of proposed 1 × 2 ETDRNA array

Achieving an antenna design for a specific frequency band (wide, notch or filter) with tunable (or changeable) centre frequency is an important challenge task. Our proposed nantenna array structure achieves this task and offers a large flexibility to the antenna in terms of operating frequency for a wide range of applications. Various parameters and design strategies such as loading slots, lumped components, switches, diodes and capacitors are introduced in order to achieve this feature. Although we have utilised a corporate feed network and the antenna geometry is based on nanoscale dimensions, which means more parametric study, but our simple and featured equilateral shape of the DR elements enable us to make natural yet simple tunability. **Figure 22** shows the rotation angles of the two equilateral triangular DR elements (T1 and T2). Since it is an equilateral triangle, the rotation angles were swept from (−180° to 180°) with a step size of 10°. The rotation angle is defined as counterclockwise movement from the bottom tip of the triangle. We apply three scenarios of rotations: (1) rotate T1 while T2 is fixed (**Figure 22a**), (2) fix T1 and rotate T2 (**Figure 22b**) and (3) simultaneously rotate T1 and T2 in the same counterclockwise direction (**Figure 22c**). In all the three scenarios, as shown in **Figure 22**, the simulation of rotation was done for the whole 360° but only five rotating angles

Figure 22. Rotation scenario of triangular DRs for spectral shifts: (a) rotate triangle T1 (left); (b) rotate triangle T2 (right) and (c) rotate both T1 and T2 simultaneously ± 90°.

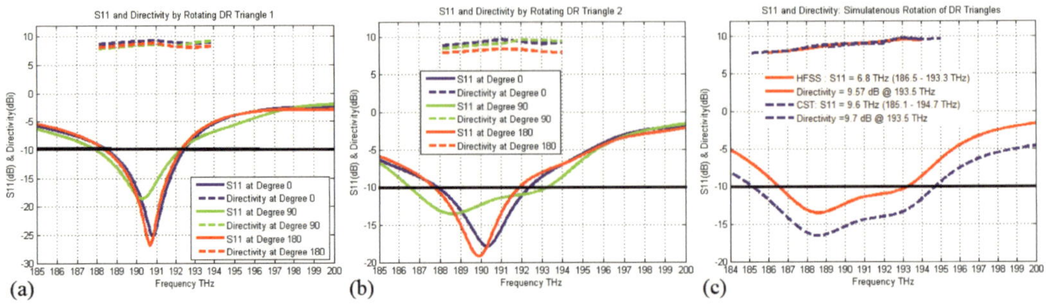

Figure 23. Return loss and directivity as per rotation: (a) T1 rotated; (b) T2 rotated and (c) both T1 and T2 simultaneously rotated.

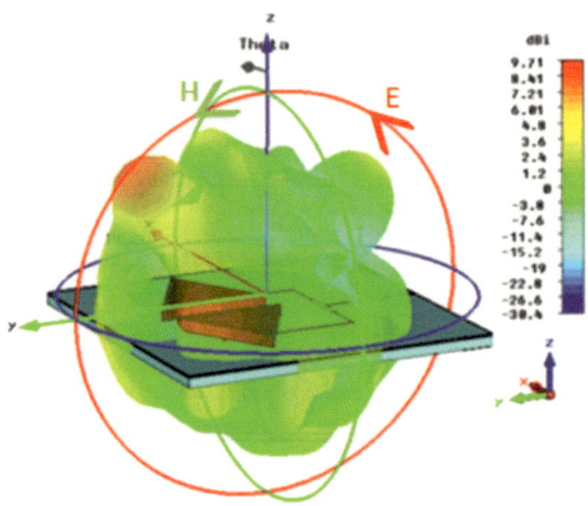

Figure 24. 3D radiation pattern of directivity at 193.5 THz after appropriate selection of DR triangular angles (−90° and 90°).

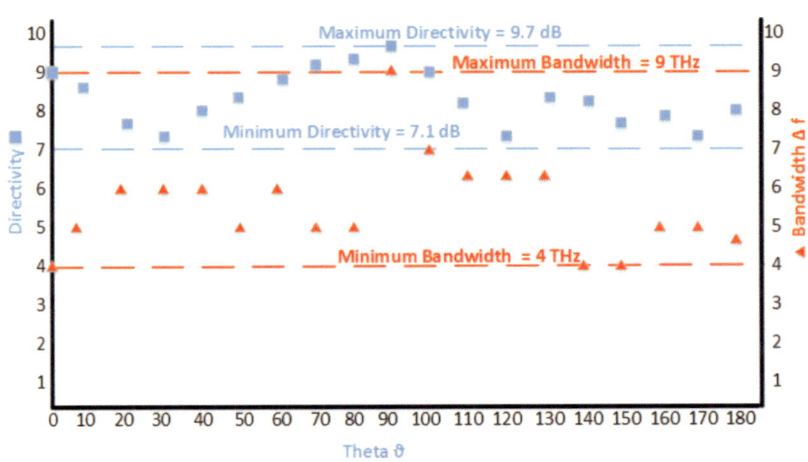

Figure 25. Effect of tunability on directivity and fractional bandwidth.

have been shown, that is, 0°, +60°, +90°, −60°, −90°, for purpose of simplicity. For our case of centre frequency at 193.5 THz, the best resonance with improved bandwidth of 4.96% (185.1–194.7 THz) and a directivity (9.7 dBi) is achieved when we opted to select the angles of T1 and T2 to be −90° and 90°, respectively. **Figure 23** displays the results in terms of s-parameters and directivity for rotation of triangles T1 and T2.

Figure 24, details of the E-field (yz-axis), shows main lobe at a direction of 45°, side lobe levels of −4.5 dB and beam width of 19°, whereas the H-field (xz-axis) has a main lobe direction of 115°, side lobe levels of −2.6 dB and beam width of 99.8° with a lot of losses in terms of high side lobe levels at a higher frequency. **Figure 25** shows the tunability effects of simulataneous rotation of triangular DRs in terms of the minimum and maximum values achieved for directivity and also the bandwidth difference Δf at −10 dB resonance in the frequency band (185–195 THz). The minimum directivity achieved is 7.15 dB and the maximum achieved is 9.71 dB when the triangles are tuned. Similary, the difference of bandwidth Δf ranges from a minimum to maximum of 4 to 9 THz, respectively.

6. Conclusion

In this chapter, we proposed two new optical nanoantenna designs working at centre frequency of 193.5 THz (1550 nm wavelength). We made use of dielectric resonators, which are ceramic based having very high dielectric permittivity, to assist the metallic designs mostly utilised in making plasmonic nanoantennas. Dielectric resonator (DR)-based antennas have made their mark and importance of antennas designed at the RF and MMW spectrum. Due to their unique feature of minimum surface loss and high radiation efficiency, they are considered nominal candidate to be used at higher spectrum bands, that is, THz.

Keeping with the offered characteristics of DR, we suggested two nanoantenna designs with triangular- and hexagonal-shaped DR materials. Both the designs are based on multilayer technology where the ground and transmission lines are made from noble metal silver. The properties of silver at optical domain are discussed and defined on the basis of Drude Model. The first design ETDRNA achieved an impedance bandwidth of 2.58% (192.3–197.3 THz), whereas the second nanoantenna design with hexagonal-shaped DR, HDLN, offered an impedance bandwidth of 3.7% (190.9–198.1 THz). Both the nanoantennas achieved high directivity of 8.6 dBi with end-fire radiation pattern. Array synthesis was also performed in order to compare and observe how much improvement is possible. The 1×2 ETDRNA achieved an improvement of 0.97 dB compared to the original directivity. Also the array antenna offers tunability in simple manner, compared to other methods discussed, making the proposed nanoantenna adaptable to many other optical frequency bands of interest for various optical communications.

Acknowledgements

This research work was supported by King Saud University, Deanship of Scientific Research and College of Engineering Center.

Author details

Waleed Tariq Sethi[1,2]*, Hamsakutty Vettikalladi[3], Habib Fathallah[2] and Mohamed Himdi[1]

*Address all correspondence to: wsethi@ksu.edu.sa

1 Institute of Electronics and Telecommunications of Rennes University (IETR), University of Rennes, France

2 KACST Technology Innovation Center in Radio Frequency and Photonics for the e-Society (RFTONICS), King Saud University, Riyadh, Saudi Arabia

3 Electrical Engineering Department, King Saud University, Riyadh, Saudi Arabia

References

[1] Cherry S. Edholm's law of bandwidth. IEEE Spectrum. 2004;**41**(7):58–60

[2] Li QC, Niu H, Papathanassiou AT, Wu G. 5G network capacity: Key elements and technologies. IEEE Vehicular Technology Magazine. 2014;**9**(1):71–78

[3] Stephan J, Brau M, Corre Y, Lostanlen Y. On the Effect of Realistic Traffic Demand Rise on LTE-A HetNet Performance. 2014 IEEE 80th Vehicular Technology Conference (VTC2014-Fall). Vancouver, BC; 2014. pp. 1–5

[4] Gregori M, Gómez-Vilardebó J, Matamoros J, Gündüz D. Wireless content caching for small cell and D2D networks. IEEE Journal on Selected Areas in Communications. 2016;**34**(5):1222–1234

[5] Mustafa IB, Uddin M, Nadeem T. Understanding The Intermittent Traffic Pattern Of HTTP Video Streaming Over Wireless Networks. 2016 14th International Symposium on Modeling and Optimization in Mobile, Ad Hoc, and Wireless Networks (WiOpt) (2016): n. pag. Web. 6 March 2017

[6] TU/e to develop 5G technology with European grant [Internet]. Available from: https://www.tue.nl/en/university/news-and-press/news/13-05-2016tueto-develop-5g-technology-with-european-grant/

[7] Valdes-Garcia A, Yong SK, Xia P. 60Ghz Technology For Gbps Wlan And Wpan. 1st ed. Hoboken, NJ: Wiley; 2013

[8] Chester E. 4.6Gbps Wi-Fi: How 60Ghz Wireless Works—And Should You Use It? Ars Technica. N.p., 2017. Web. 6 March 2017

[9] Researchers Demonstrate World's First 5G, 100 To 200 Meter Communication Link Up To 2 Gbps". Phys.org. N.p., 2017. Web. 6 March 2017

[10] Kürner T, Priebe S. Towards THz communications-status in research, standardization and regulation. Journal of Infrared, Millimeter, and Terahertz Waves. 2014;**35**(1):53–62

[11] Nagatsuma T, Ducournau G, Renaud CC. Advances in terahertz communications accelerated by photonics. Nature Photonics. 2016;**10**(6):371–379

[12] Petrov V, Pyattaev A, Moltchanov D, Koucheryavy Y. Terahertz band communications: Applications, research challenges, and standardization activities. Lisbon: 2016 8th International Congress on Ultra Modern Telecommunications and Control Systems and Workshops (ICUMT); 2016. pp. 183–190

[13] Lukasz L, Brzozowski M, Kraemer R. Data Link Layer Considerations for Future 100 Gbps Terahertz Band Transceivers. Wireless Communications and Mobile Computing. Vol. 2017. Article ID 3560521, 11 pages, 2017

[14] Yilmaz T, Akan OB. On the use of low terahertz band for 5G indoor mobile networks. Computers and Electrical Engineering. 2015;**48**:164–173

[15] Akyildiz IF, Jornet JM, Han C. Terahertz band: Next frontier for wireless communications. Physics Communications. 2014;**12**:16–32. DOI: 10.1016/j.phycom.2014.01.006 ()

[16] Obayya S, Areed NFF, Hameed MFO, Hussein M. Optical nano-antennas for energy harvesting. In: Innovative Materials and Systems for Energy Harvesting Applications. Hershey, PA, USA: IGI Global; 2015. p. 26

[17] Yang Y, Li Q, Qiu M. Broadband nanophotonic wireless links and networks using on-chip integrated plasmonic antennas. Scientific Reports. 2016;**6**:19490. DOI: 10.1038/srep19490

[18] Yousefi L, Foster AC. Waveguide-fed optical hybrid plasmonic patch nano-antenna. Optics Express. 2012;**20**(16):18326–18335

[19] Hong SG, et al. Bioinspired optical antennas: Gold plant viruses. Light: Science & Applications. 2015;**4**:267–272

[20] Horiuchi N. Optical antennas: Reconfigurable resonance. Nature Photonics. 2016;**10**(5): 2680–2685

[21] Jornet JM, Akyildiz AF. Channel modeling and capacity analysis for electromagnetic wireless nanonetworks in the terahertz band. IEEE Transactions on Wireless Communications. 2011;**10**(10):3211–3221

[22] Novotny L, van Hulst N. Antennas for light. Nature Photonics. 2011;**5**:83–90

[23] Palik ED. Handbook of Optical Constants of Solids. New York: Academic Press; 1998

[24] Johnson PB, Christy RW. Optical constants of the noble metals. Physical review B, APS Physics. 1972;**6**(12):5056–5070

[25] Petosa A, Ittipiboon A. Dielectric resonator antennas: A historical review and the current state of the art. IEEE Antennas and Propagation Magazine. 2010;**52**:91–116

[26] Luk KM, Leung KW, editors. Dielectric Resonator Antennas. Baldock, England: Research Studies Press; 2003

[27] Petosa A. Dielectric Resonator Antenna Handbook. Norwood, MA: Artech House; 2007

[28] Keyrouz and Caratelli D. Dielectric resonator antennas: Basic concepts, design guidelines, and recent developments at millimeter-wave frequencies. International Journal of Antennas and Propagation. 2016;**10**:0–20

[29] Belov PA, et al. Superdirective all-dielectric nanoantennas: Theory and experiment. IOP Conference Series: Materials Science and Engineering. 2014;**67**:012008–012013

[30] Devi I, et al. Modeling and design of all-dielectric cylindrical nanoantennas. Journal of Nanophotonics. 2016;**10**(4):046011–046011

[31] Andryieuski A, Malureanu R, Biagi G, Holmgaard T, Lavrinenko A. Compact dipole nanoantenna coupler to plasmonic slot waveguide. Optics Letters. 2012;**37**:1124–1126

[32] Suh JY, Huntington MD, Kim CH, Zhou W, Wasielewski WR, Odom TW. Extraordinary nonlinear absorption in 3d bowtie nanoantennas. Nano Letters. 2012;**12**:269–274

[33] Maksymov IS, Miroshnichenko AE, Kivshar YS. Actively tunable bistable optical yagi-uda nanoantenna. Optics Express. 2012;**20**:8929–8938

[34] Patel SK, Argyropoulos C. Plasmonic nanoantennas: Enhancing light-matter interactions at the nanoscale. EPJ Applied Metamaterials. 2015;**2**:03140–03174

[35] Giannini V, et al. Plasmonic nanoantennas: Fundamentals and their use in controlling the radiative properties of nanoemitters. Chemical Reviews. 2011;**111**(6): 3888–3912

[36] Wu CH, et al. Multimodal magneto-plasmonic nanoclusters for biomedical applications. Advanced Functional Materials. 2014;**24**(43):6862–6871

[37] Maksymov IS. Magneto-plasmonic nanoantennas: Basics and applications. Reviews in Physics. 2016;**1**:36–51

[38] Professor Nicholas Fang's Research Group @ MIT. Web.mit.edu. N.p., 2017. Web. 6 March 2017

[39] Johnson PB, Christy RW. Optical constants of the noble metals. Physical Review B. 1972;**6**(12):4370–4379

[40] Sinha R, Karabiyik M, Al-Amin C, Vabbina PK, Guney DO, Pala N. Tunable room temperature THz sources based on nonlinear mixing in a hybrid optical and THz micro-ring resonator. Scientific Reports. 2015;**5**:9422

[41] Lo HY, Leung KW, Luk KM, Yung EKN. Lowprofile equilateral-triangular dielectric resonator antenna of very high permittivity. Electronics Letters. 1999;**35**(25):2164–2166

[42] Kishk AA. A triangular dielectric resonator antenna excited by a coaxial probe. Microwave and Optical Technology Letters. 2001;**30**(5):340–341

[43] Balanis C. Antenna Theory: Analysis and Design. New York, NY,USA: John Wiley & Sons; 2005

[44] Olmon RL, Raschke MB. Antenna-load interactions at optical frequencies: Impedance matching to quantum systems. Nanotechnology. 2012;**23**(44):444001

[45] Neubrech F, et al. Resonances of individual lithographic gold nanowires in the infrared. Applied Physics Letters. 2008;**93**(16)163105

[46] Hattori HT, Li Z, Liu D, Rukhlenko ID, Premaratne M. Coupling of light from microdisk lasers into plasmonic nanoantennas. Optics Express. 2009;**17**(23):20878–20884

[47] Li Z, Hattori HT, Fu L, Tan HH, Jagadish C. Merging photonic wire lasers and nanoantennas. Journal of Lightwave Technology. 2011;**29**(18):2690–2697

[48] Hattori HT, Li Z, Liu D. Driving plasmonic nanoantennas with triangular lasers and slot waveguides. Applied Optics. 2011;**50**(16):2391–2400

[49] Hamsakutty V, Mathew KT. Hexagonal dielectric resonator antenna: A novel DR antenna for wireless communication [Ph.D. thesis]. Department of Electronics, Cochin University of Science and Technology, Dyuhti, India, 2007.

[50] Mongia RK, Bhartia P. Dielectric resonator antennas: A review and general design relations for resonant frequency and bandwidth. International Journal of Microwave and Millimeter-Wave Computer-Aided Engineering. 1994;**4**(3):230–247

Chapter 4

Improvement of the Quantum Dot-in-a-Well (QDWELL) Laser and Amplifier Performance under the Optical Injection

Yossef Ben Ezra and Boris I. Lembrikov

Additional information is available at the end of the chapter

http://dx.doi.org/10.5772/intechopen.69946

Abstract

Quantum dot (QD) laser devices can be successfully used in optical communications due to their unique properties caused by the carrier localization in three dimensions. In particular, quantum dot-in-a-well (QDWELL) lasers are characterized by an extremely low threshold current density and the high modulation frequency. However, their operation rate is limited by the strongly nonlinear electron and hole scattering rates in and out of QD. We investigated theoretically the nonlinear optical phenomena in QDWELL lasers and amplifiers under the optical injection. We have shown that the synchronization of the carrier dynamics in QD and quantum well (QW) caused by the optical injection improves the QDWELL laser performance and, in particular, enhances the relaxation oscillation (RO) frequency. As a result, the QDWELL laser performance in the analogous optical link (AOL) is significantly improving. The optical injection also improves the performance of the QDWELL-based semiconductor optical amplifiers (SOA).

Keywords: quantum dot (QD), quantum well (QW), laser, semiconductor optical amplifier (SOA), optical communications

1. Introduction

Advanced high-capacity communication systems are necessary for different applications, such as medical diagnosis, traffic safety, Internet, data services, etc. [1]. These new applications generate a giant data traffic which requires the time sensitive analysis and data processing at high-performance computing infrastructures (HPC) and the data storage, transport, and exchange in datacenters (DC) [1]. Recently, all-optical architecture of DC has been proposed.

It is based on the switching of all data in the optical domain [1]. This approach requires the development of new photonic devices [1]. One of the most efficient technologies for the realization of such devices is Silicon Photonics. Silicon Photonics can resolve the so-called bandwidth bottleneck by the integration of photonic integrated circuits (PIC) and electronic integrated circuits [1]. The active components such as lasers are made of the III-V compositions. There are two techniques of the III-V laser integration with PIC: (i) the III-V laser die can be butt-coupled to the silicon photonic chip using active alignment and (ii) III-V materials are wafer bonded to the silicon photonic chip in order to fabricate lasers lithographically aligned to the silicon waveguide circuit [1].

A generic optical communication system consists of an optical transmitter, optical communication channel, and an optical receiver [2]. A block diagram of such a system is shown in **Figure 1** [2].

The information highways providing these services are based on optical fibers [1]. Typically, silica optical fibers are used as the communication channel due to their low losses of about 0.2 dB/km [2]. In such a case, the losses are 20 dB after the propagation distance of 100 km, and the optical power decrease by 100 times defines the amplifier spacing in the long-haul lightwave systems [2, 3]. In long-haul networks extended over thousands of kilometers fiber losses are compensated by using a chain of amplifiers boosting the signal power periodically to its original value [1, 2]. Erbium-doped fiber amplifiers (EDFA) are widely used in optical communication systems due to their compatibility with transmission fibers, energy efficiency, and low cost [1, 2]. Raman amplifier operation is based on the stimulated Raman scattering (SRS) in silica fibers [2]. The advantages of the Raman amplifiers are the using of the fiber itself as an active medium and the large bandwidth [2]. However, the Raman amplifiers require a comparatively large pumping power [2].

The optical communication system performance is limited by the fiber dispersion leading to the optical pulse broadening with propagation through the channel [2, 3]. As a result, the original signal recovery with the high enough accuracy may be impossible [2]. The dispersion influence is strongly manifested in multimode fibers (MMFs) where different fiber modes propagate with different velocities. For this reason, optical communication systems are mainly based on single-mode fibers (SMFs) [2]. In such a case, the intermodal dispersion vanishes because the pulse energy is carried by a single mode [2]. The main types of the SMF dispersion are as follows: (i) group velocity dispersion (GVD) is caused by group velocity v_g frequency dependence (chromatic dispersion); (ii) material dispersion is caused by the silica refractive index n dependence on the optical frequency; (iii) waveguide dispersion is defined by the following parameters: the fiber

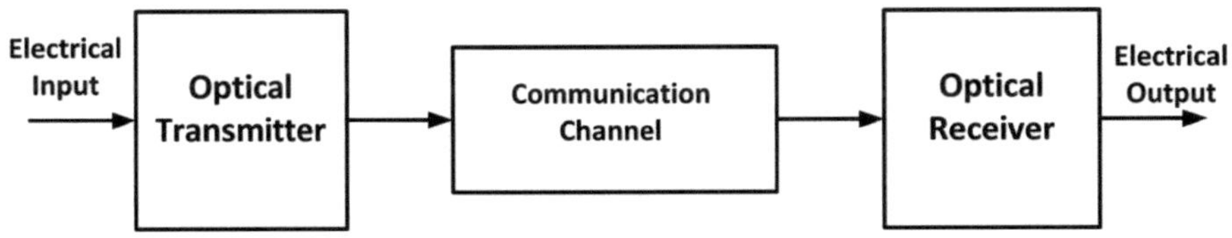

Figure 1. A block diagram of a generic optical communication system.

core radius r_0, the fiber core refractive index n_1 and the cladding refractive index n_2; and (iv) polarization-mode dispersion (PMD) is caused by a modal birefringence [2, 3]. SMF material dispersion is small enough, and it actually limits the optical communication system bit rate and transmission distance under the condition that other types of dispersion are mitigated [2].

Optical transmitter converts the electrical signal into the optical signal and launches it into the optical fiber [2]. An optical transmitter consists of an optical source, a modulator, and a channel coupler [2]. An optical source is usually a semiconductor laser or light-emitting diode (LED) compatible with the optical communication channel [2]. In LED, population inversion is not realized, and the light is generated through spontaneous emission caused by the radiative recombination of electron-hole pairs in the active layer [2]. In semiconductor lasers, the stimulated emission of light is the dominating operation mechanism [2, 4]. For this reason, the LED radiation is rather weak as compared to the semiconductor laser light. The typical values of the launched power are less than 100 μW (−10 dBm) for LED and about 1 mW (10 dBm) for semiconductor lasers [2]. The optical signal is generated by the direct or external modulation of the optical carrier wave radiated by a semiconductor laser [2]. Generally, time, quadrature, polarization, and frequency are widely used in optical networking technologies for complex quadrature modulation formats, polarization multiplexing, digital pulse shaping, and coherent detection [1]. Recently, space as a physical dimension for modulation and multiplexing in communication systems attracted interest for fiber capacity scaling [1]. Space-division multiplexing (SDM) uses multiplicity of space channels, or spatial parallelism, in order to increase capacity of the optical communication system [1]. For instance, fiber bundles 10 × 10 Gb/s or 4 × 25 Gb/s can be used for the implementation of 100 Gb/s commercial client interfaces [1].

In the case of the direct modulation, the semiconductor laser is biased near the threshold and driven by the electrical sinusoidal signal for analog modulation or electrical signal bit stream for digital modulation [2]. In the case of the external modulation, two types of the external modulators are mainly used: the Mach-Zehnder modulator (MZM) and the electro absorption modulator (EAM) [2, 4]. The channel coupler is a microlens focusing the optical signal onto the optical fiber entrance plane [2]. The bit rate of optical transmitters is limited by electronic components [2].

An optical receiver converts the optical signal into the electrical domain and recovers the transmitted data [2, 4]. The structure of the optical receiver depends on the modulation format. Consider first the on-off keying (OOK) modulation, where an electrical binary stream modulates the optical carrier intensity inside an optical transmitter [2, 4]. In such a case, the directly modulated optical signal after the propagation through the optical fiber is converted in the receiver directly into the original digital signal in the electrical domain [2]. The main component of the receiver is a semiconductor photodetector (PD) converting light into electrical signal through the photoelectric effect [2]. Such a communication system is intensity modulation with direct detection (IM/DD) system [2]. PD should possess high sensitivity, fast response, low noise, low cost, and high reliability [2]. Generally, the transmission quality is characterized by the received signal-to-noise ratio (SNR) given by $SNR = 20log_{10}(A_{max}/A_N)$ dB, where A_{max} and A_N are the maximum amplitude of the given analog signal and the root mean square (RMS) noise amplitude of the analog signal, respectively [2]. The total noise is a stochastic process consisting of a number of components such as mode partition noise (MSN), laser intensity noise, modal noise, shot noise, thermal noise, dark current noise, amplified spontaneous emission (ASE), and crosstalk noise [4]. The shot noise and thermal noise are the two fundamental noise mechanisms responsible for the current fluctuations in optical receivers [2]. The detailed analysis of the different types of noise may be found in Refs. [2–4]. Here, we should note that the thermal noise limits PD performance in practice, whereas the shot noise determines a fundamental limit of PD operation [2]. For instance, in the shot-noise limit, SNR $= 20$ dB can be realized for PD quantum efficiency close to unity and a number of photons $N_p = 100$ in the input optical signal [2].

Unlike the IM/DD system, the performance of the system with a coherent detection technique is limited by the shot-noise alone [2, 4, 5]. Another important advantage of the coherent detection is the possibility of the detection of signals with advanced modulated formats such as frequency-shift keying (FSK), binary phase-shift keying (PSK), quadrature PSK, and 16-quadrature amplitude modulation (QAM) allowing bit rates of 50, 100, and 200 Gb/s [1, 2, 4]. In the case of the coherent detection, the input optical signal with the optical carrier frequency ω_0 is mixed with the continuous wave (CW) optical signal generated at the receiver by a narrow linewidth laser called the local oscillator (LO) with the frequency ω_{LO} [2]. The output contains the signal with the intermediate frequency $\omega_{IF} = \omega_0 - \omega_{LO} \ll \omega_0$ which falls on PD [2]. There are two types of the coherent detection technique: (i) a homodyne detection when $\omega_0 = \omega_{LO}\,; \omega_{IF} = 0$; (ii) a heterodyne detection when $\omega_0 \neq \omega_{LO}$ [2, 4]. In the latter case, the detection of the optical signal is carried out in two stages: first, the carrier frequency ω_0 is converted to the intermediate frequency ω_{IF} usually belonging to the radio frequency (RF) range; then, the RF signal is down converted to the modulating signal bit stream [2, 4].

Optical signal processing in optical communication systems is based on the linear and nonlinear optical techniques used for the manipulation and processing of digital, analogue, and quantum information [1]. Ultrafast optical nonlinearities provide an operation rate advantage as compared to the electronic techniques for switching, regeneration, wavelength conversion, performance monitoring, and analog digital conversion (ADC) [1]. In particular, semiconductor optical amplifiers (SOAs) can be used in such applications due to their high operation rate and strong nonlinearity [1–3].

The brief review of the optical communication systems clearly shows that a semiconductor laser is a key component of both the transmitter and receiver. Evidently, for the highly efficient applications in optical communication systems, semiconductor laser performance should be characterized by low-threshold current, high-speed direct modulation, ultrashort optical pulse generation, narrow spectral linewidth, broad modulation bandwidth, comparatively high optical output power, low relative intensity noise (RIN), low cost, and low electrical power consumption [2, 6]. The objective of this chapter is the investigation of the optical injection influence on the performance of the novel semiconductor laser based on quantum dots (QD) in a quantum well (QW) structure. The number of publications concerning semiconductor lasers in general and QD lasers in particular is enormous and hardly observable. For this reason, in Section 2, we briefly discuss the structure, operation principle, and basic characteristics of a semiconductor laser widely used in optical communication systems. In Section 3, the fundamentals of a QD in QW (QDWELL) laser and SOA are discussed. The original theoretical

http://dx.doi.org/10.5772/intechopen.69946

results related to the dynamics and performance of the optically injected QDWELL lasers and SOA are presented in Section 4. The conclusions are presented in Section 5.

2. Structure and operation principle of a semiconductor laser

Semiconductor laser is an electrically pumped p-i-n diode [6]. In the forward-biased diode, electrons in the conduction band and holes in the valence band are injected into the active layer placed between the p-type and n-type cladding layers [2, 6]. The structure of an edge-emitting semiconductor laser is shown schematically in **Figure 2** [2].

The cladding layers are made of the semiconductor material with the bandgap E_{g2} larger than the active layer bandgap E_{g1} [2, 6]. Such a structure consisting of the p-n junctions made of semiconductor materials with different bandgaps is called a heterostructure [2, 7]. As a result, the electrons and holes are confined in the active layer.

The $Al_xGa_{1-x}As/GaAs/Al_xGa_{1-x}As$ heterostructure is shown in **Figure 3** [6]. The concentration of donors N_D and acceptors N_A in the n-type and p-type regions, respectively, should satisfy the condition that the separation $E_{FC} - E_{FV}$ between the electron Fermi energy level E_{FC} and the hole Fermi energy level E_{FV} is larger than the bandgap energy E_g [2]. Recombination of the electron-hole pair in the presence of the optical field results in the photon stimulated emission [2, 6].

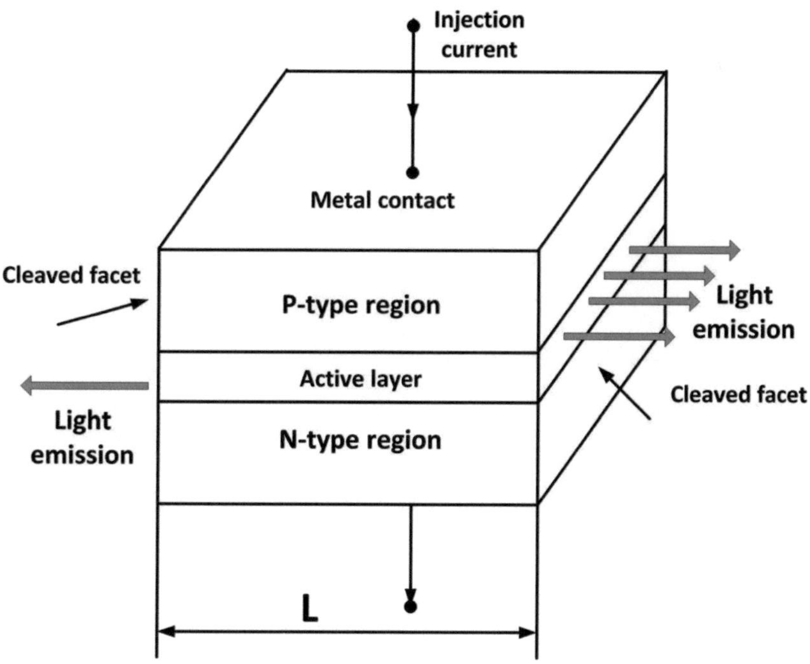

Figure 2. Edge-emitting semiconductor laser.

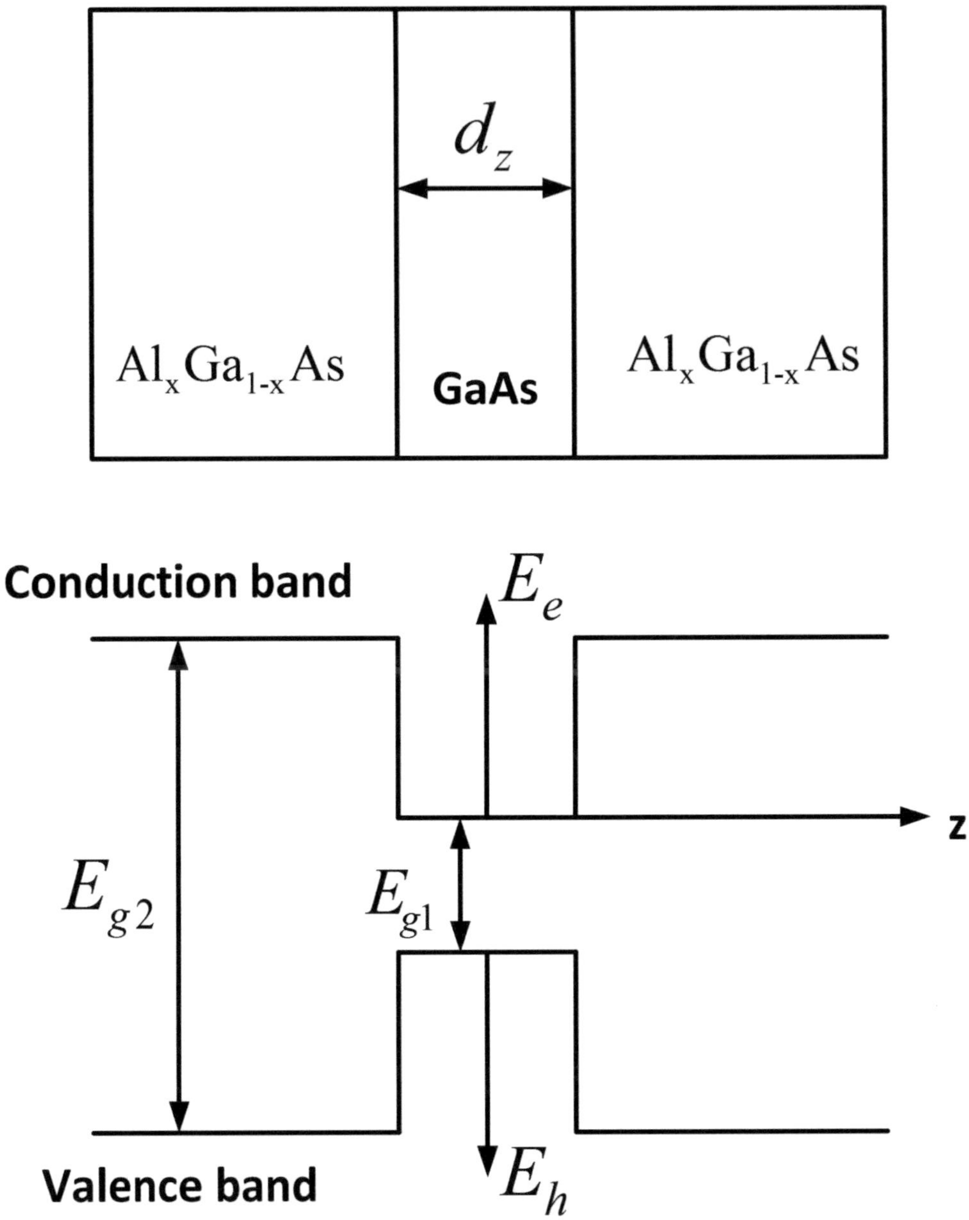

Figure 3. The energy bands of the semiconductor laser based on the $Al_xGa_{1-x}As/GaAs/Al_xGa_{1-x}As$ heterostructure.

Usually, the active layer has a larger refractive index than the cladding layers providing also the confinement of the photons generated in the active layer [2]. When the injected carrier density in the active layer becomes larger than a critical value, the population inversion occurs and the active region manifests an optical gain [2]. Input signal intensity P_{in} in such a case is amplified, and the output signal intensity P_{out} has the form [2]

$$P_{out}(L) = P_{in}(0)\exp(gL) \tag{1}$$

where g is the power gain coefficient and L is the active layer length. Usually, the laser dimensions are less than 1 mm in all three directions which result in compact design [2]. The gain is negative $g < 0$ for the injected carrier density N below the population inversion value. The gain becomes positive and increases with the increase of the injected carrier density N. The peak gain g_p is given by the approximate empirical relationship [2].

$$g_p(N) = \sigma_g(N - N_T) \tag{2}$$

where σ_g is the gain cross section or the differential gain and N_T is the transparency value of the carrier density. For InGaAsP lasers, the typical values of these parameters are following: $\sigma_g \approx (2-3) \times 10^{-16}$ cm^{-2}, $N_T \approx (1-1.5) \times 10^{18}$ cm^{-3} [2]. Eq. (2) can be used in the high-gain region where g_p >100 cm^{-1} [2]. The laser operation also requires an optical feedback provided by placing the gain medium into a Fabry-Perot (FP) cavity formed in our case by the two cleaved laser facets serving as mirrors [2]. In such a case, a critical value of the gain g_{cr} is given by [2]

$$g_{cr} = \alpha_{\text{int}} + \frac{1}{2L}\ln\left(\frac{1}{R_1R_2}\right) \tag{3}$$

where α_{int} is the coefficient of the internal losses caused by free carrier absorption and interface scattering and $R_{1,2}$ are the facet reflectivities. In the laser FP cavity, only the longitudinal modes can exist determined by the condition [2]

$$2\beta L = m\pi;\ \beta = \frac{2\pi\nu n_{\text{mode}}}{c};\ \nu = \nu_m = \frac{mc}{2n_{\text{mode}}L};\ m = 1, 2, \ldots \tag{4}$$

where β, n_{mode}, and c are the mode propagation constant, the mode refractive index, and the free space light velocity, respectively. Eq. (4) shows that the laser frequency ν corresponds to the discrete spectrum of the laser longitudinal modes. A semiconductor laser radiation typically consists of several longitudinal modes with a gain spectrum bandwidth of about 10 THz [2]. The dominant mode is the one with gain closest to the peak gain g_p [2]. The optical modes penetrate into the cladding layers due to the boundary conditions for the optical mode electric and magnetic field while the gain takes place only in the active layer. Hence, the confinement factor of the active layer $\Gamma = g/g_m$ is introduced which is typically less than 0.4 [2]. Here, g_m is the so-called material gain. The modes propagate inside the optical fiber with different speeds because of GVD as it was mentioned above. Consequently, the multimode semiconductor lasers are limiting the optical communication operation rate [2].

There exist single-mode semiconductor lasers emitting light mainly in a single longitudinal mode such as distributed feedback (DFB) lasers, coupled-cavity lasers, tunable lasers, and vertical-cavity surface-emitting lasers (VCSEL) [2]. The detailed analysis of structure and operation principles of these lasers can be found in Ref. [2] and references therein. Here, we only briefly discuss the peculiarities of these lasers following Ref. [2]. DFB laser has a built-in periodic grating with the period d_0 which results in refraction index periodicity. Consequently, the waves propagating in the forward and backward directions in the cavity are coupling, and the Bragg diffraction occurs. The feedback in such a case is distributed throughout the laser cavity length. The coupling occurs only for the modes with the Bragg wavelength λ_B satisfying the Bragg condition $d_0 = (\lambda_B/2n_{\text{mode}})m, m = 1, 2, 3, \ldots$ [2]. This condition provides the mode selectivity of the DFB lasers. The strongest coupling between the forward and backward propagating modes occurs for $m = 1$ [2].

In a coupled-cavity semiconductor laser, single-mode operation is realized by coupling the laser cavity to an external cavity. The in-phase feedback occurs only for the laser modes with a wavelength which almost coincides with a certain longitudinal mode of the external cavity [2]. The effective reflectivity of the laser facet close to the external cavity depends on the wavelength, and the losses of certain modes sharply decrease. The longitudinal mode which simultaneously has the highest gain and the lowest cavity loss becomes the dominant mode [2].

A tunable semiconductor laser typically consists of three sections: the active section, the phase-control section, and the Bragg section [2]. Each one of these sections is biased independently by injecting different currents in a following way: the current injected into the Bragg section changes the Bragg wavelength λ_B through carrier induced variations in the refractive index n; the current injected into the phase-control section changes the feedback phase through carrier induced variations in the refractive index of this section [2]. As a result, the laser wavelength can be tuned over the range of 10–15 nm by controlling the injection current in the phase-control and the Bragg sections [2].

VCSEL emits light in a direction perpendicular to the active layer plane and operates in a single longitudinal mode regime due to the small cavity length of about 1 μm in a wide range of wavelengths of about 650–1600 nm [2]. The emitted light has a form of a circular beam which can be inserted into SMF with high efficiency [2].

The dynamics of the single-mode semiconductor laser is described by the following rate equations for the number of electrons N and photons P in the active layer [2].

$$\frac{dP}{dt} = GP + R_{sp} - \frac{P}{\tau_p} \tag{5}$$

$$\frac{dN}{dt} = \frac{I}{q} - \frac{N}{\tau_c} + GP \tag{6}$$

where I is the injection current, q is the electron charge, $G = (\Gamma v_g \sigma_g / V)(N - N_0)$ is net rate of the stimulated emission, V is the active layer volume, $N_0 = N_T V$, v_g is the light group velocity, R_{sp} is the spontaneous emission rate, and τ_p are τ_c are the photon and carrier lifetime, respectively. The threshold current I_{th} for the laser generation regime obtained from the steady-state solution of Eqs. (5) and (6) has the form [2].

$$I_{th} = \frac{q}{\tau_c}\left(N_0 + \frac{V}{G_N \tau_p}\right); \qquad G_N = \frac{\Gamma v_g \sigma_g}{V} \tag{7}$$

The direct modulation of the laser emission can be realized if the injection current includes a time-dependent component. In the case of a sinusoidal modulation, the modulation frequency ω_m is limited by the relaxation oscillation (RO) frequency Ω_R given by [2]

$$\Omega_R = \left[G G_N P_b - \frac{(\Gamma_P - \Gamma_N)^2}{4} \right]^{1/2} \tag{8}$$

where P_b is the steady-state photon number, $\Gamma_P = \left(R_{sp}/P_b\right) + \varepsilon_{NL} G P_b; \Gamma_N = \tau_c^{-1} + G_N P_b,$

ε_{NL} is a nonlinear gain parameter, and $\Gamma_R = (\Gamma_N + \Gamma_P)/2$ is a damping rate of the relaxation oscillations. The efficiency of the laser modulation is significantly reduced for $\omega_m > \Omega_R$ [2]. The 3 dB modulation bandwidth f_{3dB} of the semiconductor laser is also determined by the RO frequency Ω_R. For $\Omega_R \gg \Gamma_{R,}$ it has the form $f_{3dB} \approx \left(\sqrt{3}\Omega_R\right)/2\pi$ [2].

Consider now QW semiconductor lasers based on a heterostructure shown in **Figure 3**. If the thickness of the active layer d_z is less than approximately 10 nm, the confined electrons and holes exhibit quantum effects, and such a heterostructure is called a single QW [6, 7]. In such a case, the De Broglie electron wavelength $\lambda = h/p$ in a semiconductor is close to the active layer thickness [7]. Here, h, p are the Planck constant and electron momentum, respectively. The charge carriers in a QW laser are localized in the z direction perpendicular to the active layer plane and represent a two-dimensional (2D) electron and hole gas [6]. The solution of the Schrödinger equation for carriers in QW produces discrete states of energy E_n in the z direction perpendicular to the QW plane and a continuous range of allowed energies in the QW plane [7]. The electron energy E_e in a QW is given by [6, 7]

$$E_e = E_n + \frac{\hbar^2 \left(k_x^2 + k_y^2\right)}{2m_e^*} \tag{9}$$

where $\hbar = h/2\pi$, m_e^* is the electron effective mass, and $k_{x,y}$ are the electron continuous momentum components in the QW plane [6, 7]. Similar expression can be written for the hole energy in QW. The threshold current in a 2D structure reduces significantly as compared to the bulk semiconductor laser [6]. The electronic and optical properties of a semiconductor laser strongly depend on the carrier density of states (DOS) $\rho_{e,h}(E_{e,h})$.The electron DOS in a QW with n_e energy levels has the form [6, 7]

$$\rho_e(E_e) = \sum_{n_e} \frac{m_e^*}{\pi \hbar^2} u(E_e - E_n); u(x) = \begin{cases} 1, x > 0 \\ 0, x < 0 \end{cases} \tag{10}$$

Similar expression can be written for the hole DOS. It is seen from Eq. (10) that carrier DOS in QW does not depend on the energy. The static and dynamic properties of the QW semiconductor laser can be described by the rate equations for the carrier density in QW N_W and in a separate confinement heterostructure (SCH) layer N_B, and the photon density in the cavity P given by [8]

$$\frac{dN_B}{dt} = \frac{I}{qV_{SCH}} - \frac{N_B}{\tau_r} - \frac{N_B}{\tau_{nb}} + \frac{N_W V_W}{\tau_e V_{SCH}} \tag{11}$$

$$\frac{dN_W}{dt} = \frac{N_B V_{SCH}}{\tau_r V_W} - \frac{N_W}{\tau_n} - \frac{N_W}{\tau_{nr}} - \frac{N_W}{\tau_e} - \frac{v_g G(N_W) P}{1 + \varepsilon P} \tag{12}$$

$$\frac{dP}{dt} = \frac{\Gamma v_g G(N_W) P}{1 + \varepsilon P} - \frac{P}{\tau_p} + \Gamma \beta_{Sp} \frac{N_W}{\tau_n} \tag{13}$$

where τ_n, τ_{nr}, τ_{nb}, τ_r, and τ_e are the radiative recombination lifetime in QW, the nonradiative recombination lifetime in QW, the total recombination lifetime in the SCH region, the carrier transport time across the SCH to QW active region, and the carrier thermionic emission time from QW into the barrier and SCH layers, respectively; β_{Sp} is the spontaneous emission feedback factor, $G(N_W)$ is the carrier density dependent gain, ε is the intrinsic gain compression factor, and V_W and V_{SCH} are the QW and SCH volumes, respectively. The rate equations for QW laser (11)–(13) are different from the bulk laser Eqs. (5) and (6). The additional Eq. (11) includes the effects of the carrier transport in SCH region typically used in QW lasers for the carrier and optical confinement [6, 8].

3. Structure and theoretical model of a quantum dot-in-a-well (QDWELL) laser and SOA

QD is a nanostructure where the electron and hole motion is confined in three dimensions reducing the degrees of freedom to zero [7, 9]. The III-V QDs are grown epitaxially on a semiconductor substrate [9]. The spontaneous formation of 3D islands during strained layer epitaxial growth is known as the Stranski-Krastanov process [9]. A continuous film of a QW thickness beneath the QD is called the wetting layer (WL) [9, 10]. In QD, carriers occupy a finite number of energy levels defined by the solution of the Schrödinger equation for a 3D potential separating the inside of QD from the outside [7]. Typically, the spherical and pyramidal QD models are used [7]. The pyramidal model is more realistic since the QD grown by means of the Stranski-Krastanov technology have approximately pyramidal form [9]. In such a case, the Schrödinger equation for QD can be solved numerically [7, 9]. Carrier DOS in QD is a $\delta-$ function for each energy level [7]. QD has a lateral size of approximately 15–18 nm, a height of about 4–5 nm, and contains several hundreds of thousands of atoms [9, 10]. The QD density per unit area in each layer is about $10^{10} - 10^{11}$ cm^{-2} [9, 10]. QD lasers exhibit extremely low threshold current due to the 3D carrier confinement [9, 10]. The spectral bandwidth of QD lasers is broader than the one of the QW lasers due to the inhomogeneous spectral broadening caused by QD size variations that exhibit a Gaussian size distribution and consequently a nearly Gaussian distribution of the emitted light frequencies [9]. Semiconductor lasers based on self-organized InGaN, InAs, InGaAlP, and In(Ga)As QD are promising candidates for applications in optical communication systems because of their low threshold current, high temperature stability, fast carrier dynamics, and the possibility of the light emission from ultraviolet to the far infrared spectral regime [9–13]. Typically, the active area of a QDWELL laser consists of several QW layers each one of a height of 4 nm with the embedded 3D confined QD having an average size of approximately 4 nm × 18 nm × 18 nm [10]. An electric current is injected into the QW layers which form the carrier reservoir (WL) where the carrier-carrier scattering leads to a filling or depletion of the confined QD levels [10]. The energy band structure of a QDWELL laser is shown in **Figure 4**. Inside the QD, the lower level or the ground state (GS) and several higher levels or excited states (ES) for electrons and holes may exist [11]. We discuss the QDWELL laser model with the GS and one ES where the emission occurs form GS [10]. In QDWELL lasers, the Coulomb interaction is the essential mechanism of the carrier-carrier scattering for high carrier densities in WL [10].

The turn-on dynamics, the small signal and large signal responses of a QDWELL laser have been studied theoretically and experimentally in the fundamental works of Lüdge, Schöll and co-workers [10, 13–18].

It has been shown that the complicated carrier dynamics in a QDWELL laser is determined by the nonradiative carrier-carrier scattering processes between the QD and the QW [17]. The

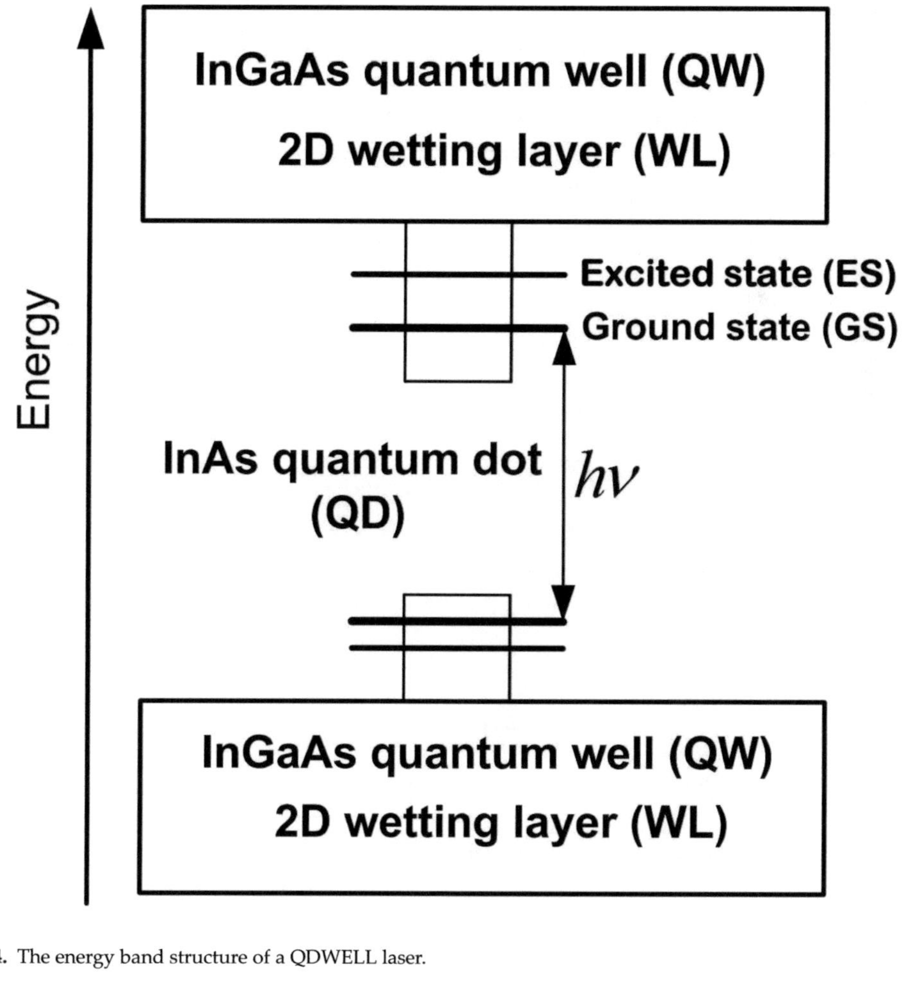

Figure 4. The energy band structure of a QDWELL laser.

corresponding scattering rates S_e^{in}, S_e^{out}, S_h^{in}, S_h^{out} of electrons (e) and holes (h) for scattering in and out of the QD are strongly nonlinear and depend on the electron and hole occupation probabilities in the confined QD levels ρ_e, ρ_h and the electron and hole densities in the QW w_e, w_h [13–18]. The electron and hole scattering times are given by, respectively, $\tau_e = \left(S_e^{in} + S_e^{out}\right)^{-1}$ and $\tau_h = \left(S_h^{in} + S_h^{out}\right)^{-1}$ [14, 16, 18]. The carrier relaxation processes in QD states are much faster than the capture processes from the QW into the QD at high carrier densities in the QW WL, and the influence of the higher QD levels can be adiabatically eliminated [17].

The nonlinearity of the scattering rates S_e^{in}, S_e^{out}, S_h^{in}, S_h^{out} and the separate dynamics of holes and electrons are essential for the QDWELL laser dynamics adequate description [16]. The electron and hole dynamics in the QDWELL laser is desynchronized due to different effective masses and different energy separation between 2D WL band edge and confined QD level [16]. The QD hole density is changing more slowly as compared to the electron density [16]. Desynchronized carrier dynamics results in the deterioration of the laser performance both in the small-signal and in the large-signal regimes [15, 17]. In particular, the QDWELL laser modulation frequency is limited by the RO frequency of 7 GHz [17]. The relatively slow carrier scattering processes from 2D WL carrier reservoir into confined QD levels are responsible for the high damping of the turn-on process and the flat modulation response curve of QDWELL lasers [10]. Usually, the modulation frequency can be increased by increasing the injection current, but in our case, the injection current increase results in the device heating and the further reduction of the RO frequency [17].

The modulation characteristics of semiconductor lasers can be improved by the optical injection locking (OIL) [19]. OIL of semiconductor lasers provides a single-mode regime and near-single-sideband modulation, strongly enhances the RO frequency and bandwidth, reduces nonlinearity, reduces RIN, reduces chirp, and increases link gain [19]. The directly modulated OIL transmission-style system combining the master laser and the slave QDWELL laser in the transmission style is shown in **Figure 5** [19].

In the transmission-style system, the injected light from the master laser enters one slave QDWELL laser facet and output is taken from the other facet [19]. The direct modulation signal is applied to the QDWELL slave laser. An isolator prevents light coupling back to the master laser. A polarization controller (PC) is used to match the overlap of master and QDWELL slave laser polarizations. The master laser light coherently combines with the QDWELL slave laser

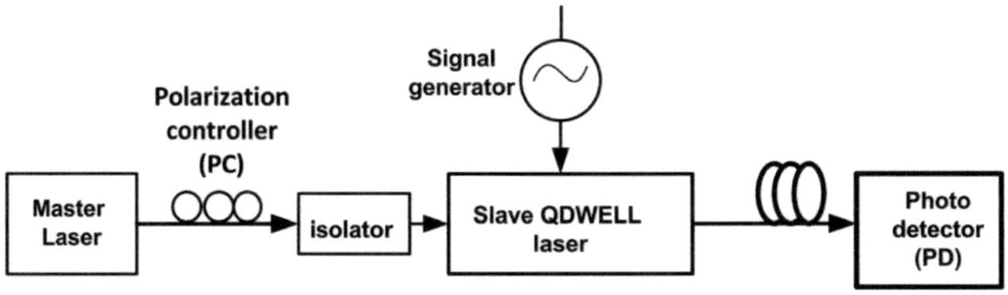

Figure 5. Directly modulated OIL system.

light changing its internal field [19]. The slave laser radiation wavelength tends to the master laser radiation wavelength until the both wavelengths become equal locking the master laser light frequency and phase [19].

The dynamics of the optically injected QDWELL laser has been studied theoretically by using the Lüdge-Schöll (LS) rate equations for the photon density per unit area n_{ph} and the phase Φ of the slave QDWELL laser radiation, the QD electron and hole occupation probabilities ρ_e, ρ_h, and the electron and hole densities in the QW w_e, w_h [20].

$$\frac{dn_{ph}}{dt} = n_{ph}\left[2\tilde{W}Z_a^{QD}(\rho_e + \rho_h - 1) - 2\kappa\right] + \frac{\beta_{Sp}}{A}2Z_a^{QD}R_{sp}(\rho_e, \rho_h) + \frac{2K}{\tau_{in}}\sqrt{n_{ph}n_{ph}^0}\cos(\Phi - 2\pi\Delta\nu_{inj}t) \tag{14}$$

$$\frac{d\Phi}{dt} = \frac{\alpha}{2}\left[2\tilde{W}(\rho_e + \rho_h - 1) - 2\kappa\right] + \frac{2K}{\tau_{in}}\sqrt{\frac{n_{ph}^0}{n_{ph}}}\sin(\Phi - 2\pi\Delta\nu_{inj}t) \tag{15}$$

$$\frac{d\rho_{e,h}}{dt} = -\tilde{W}A(\rho_e + \rho_h - 1)n_{ph} - R_{sp}(\rho_e, \rho_h) + S_{e,h}^{in}(w_e, w_h)\left(1 - \rho_{e,h}\right) - S_{e,h}^{out}(w_e, w_h)\rho_{e,h} \tag{16}$$

$$\frac{dw_{e,h}}{dt} = \frac{j}{q} - 2N^{QD}\left[S_{e,h}^{in}(w_e, w_h)\left(1 - \rho_{e,h}\right) - S_{e,h}^{out}(w_e, w_h)\rho_{e,h}\right] - \tilde{R}_{sp} \tag{17}$$

Here, A is the in-plane area of the QW, ν_L and ν_{inj} are the close frequencies of the QDWELL laser and the injected light, respectively, $K = \sqrt{T_{inj}n_{inj}/n_{ph}^0}$ is the injection strength, T_{inj} is the transmission coefficient of the cavity mirror, n_{inj} is the injected photon density in the QDWELL laser active region, and n_{ph}^0 is the steady-state photon density without injection; $\Delta\nu_{inj} = \nu_{inj} - \nu_L$ is the input detuning, $Z_a^{QD} = a_L A N_a^{QD}$ is the number of active QDs inside the waveguide, a_L is a number of self-organized QD layers, N_a^{QD} is the density per unit area of the active QD of a lasing subgroup, N^{QD} is the density per unit area of all QDs; 2κ are the optical intensity losses; $\tilde{W}$, and W are the Einstein coefficients for the coherent and incoherent interaction of the QD with all resonator modes, respectively; α is the linewidth enhancement factor (LEF), β_{Sp} is the spontaneous emission factor, $R_{sp}(\rho_e, \rho_h)$, and $\tilde{R}_{sp}$ are the spontaneous emission rates in QD and QW, respectively; τ_{in} is the time of one round trip of the light in the cavity of length L. We do not present here the complicated explicit expressions of the scattering rates S_e^{in}, S_e^{out}, S_h^{in}, S_h^{out} which can be found in Ref. [20]. It has been shown that the QDWELL laser's dynamic properties under optical injection are strongly influenced by the Coulomb scattering processes which can be taken into account by using the microscopically calculated carrier lifetimes $\tau_{e,h}$ [20]. The nonlinear dynamics of the optically injected QDWELL laser is determined by the injection strength and the input detuning [20]. The frequency-locking range depends on the pump current. The photon density as well as the electron and hole densities in QD and QW increase with increasing the pump current which results in the change of scattering rates and the QDWELL turn-on behavior [20]. The QDWELL laser under the optical injection manifests complex oscillatory dynamics for a comparatively large detuning $\Delta\nu_{inj}$ [20]. The large regions of stable CW operation occur for the short carrier lifetimes of about 10 ps typical for strongly damped relaxation oscillations [20].

Under certain conditions, QDWELL laser can operate as a SOA for high-speed applications [9]. Generally, SOA is the LED amplifying the input optical signal. SOAs are divided into two groups depending on their structure: (i) traveling wave amplifiers (TWA) and (ii) FP cavity amplifiers [9]. The block diagram of the TWA QDWELL SOA is shown in **Figure 6**.

SOA attracted a wide interest for applications in optical communication systems due to their small size, high gain, strong nonlinearity, and the possibility of on-chip optoelectronic integration [2, 9]. QD SOAs are particularly promising candidates for the amplification of ultrafast pulses in the relatively broad spectral range due to their ultrafast carrier dynamics and inhomogeneous spectral broadening [9]. The SOA dynamics is described by the carrier rate equations, equation for the time-dependent SOA gain and the equations for the slowly varying envelopes (SVE) of the optical pulses [2, 9].

The system of rate equations (14)–(16) should be modified for the case of a TWA QDWELL SOA. Eq. (17) for the electron and hole densities in the QW w_e, w_h does not change. The modified equations for the pumping and signal photon densities per unit area $n_{ph,p,s}$, phases $\Phi_{p,s}$, and the QD electron and hole occupation probabilities ρ_e, ρ_h in QDWELL SOA have the form [20–23].

$$\frac{\partial n_{ph,p,s}(z,\tau)}{\partial z} = \left(g_{\text{mod }p,s} - 2\kappa\right) n_{ph,p,s}(z,\tau) \tag{18a}$$

$$\frac{\partial \Phi_{p,s}}{\partial z} = -\frac{\alpha}{2} g_{\text{mod }p,s} \tag{18b}$$

$$\frac{d\rho_{e,h}}{dt} = -\tilde{W}A\left(\rho_e + \rho_h - 1\right)\left(n_{ph,p} + n_{ph,s}\right) - R_{sp}\left(\rho_e, \rho_h\right) + S_{e,h}^{in}(w_e, w_h)\left(1 - \rho_{e,h}\right) - S_{e,h}^{out}(w_e, w_h)\rho_{e,h} \tag{19}$$

where $\tau = t - z/v_g$, and $g_{modp,s}$ is the modal gain [22].

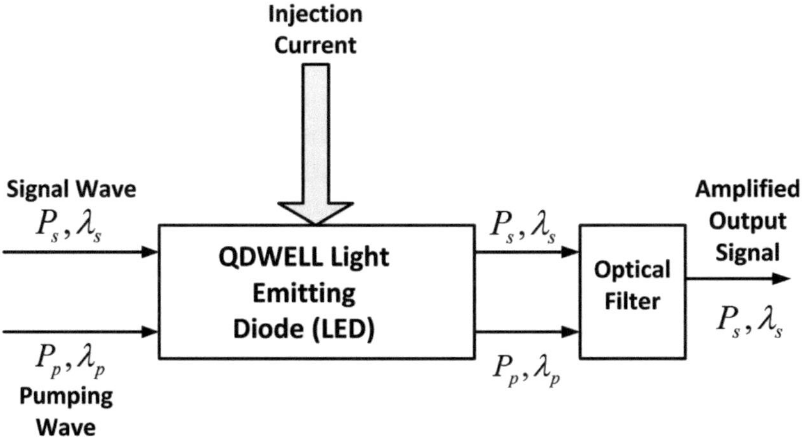

Figure 6. The block diagram of TWA QDWELL SOA.

$$g_{\text{mod}} = \frac{2\Gamma}{r_{QD}} N^{QD} \int d\omega_i F(\omega_i) \sigma_i(\omega_0) \left(\rho_e + \rho_h - 1\right) \quad (20)$$

Here, r_{QD} is the mean size of QD, $\sigma_i(\omega_0)$ is the cross section of interaction of photons of frequency ω_0 with carriers in QD at the transition frequency ω_i:

$$\sigma_i(\omega_0) = \frac{\sigma_{res}}{1 + (\omega_i - \omega_0)^2 T_2^2} \quad (21)$$

where σ_{res} is the resonant cross section, $T_2 = 2\gamma_{hom}^{-1}$ is the dephasing time, and γ_{hom} is the homogeneous full width of half maximum (FWHM). The distribution function $F(\omega_i)$ of the transition frequency in the QD ensemble describes the inhomogeneous broadening caused by the QD size variation. It is assumed to be Gaussian and has the form [22]

$$F(\omega_i) = \frac{1}{\Delta\omega\sqrt{\pi}} \exp\left[-\frac{(\omega_i - \overline{\omega})^2}{(\Delta\omega)^2}\right] \quad (22)$$

where $\overline{\omega}$ is the average transition frequency, $\Delta\omega = \gamma_{inhom}(2\sqrt{\ln 2})^{-2}$, and γ_{inhom} is the inhomogeneous FWHM.

QD SOA can be used for the ultrafast and distortion-free amplification at pulse repetition rates of 20, 40, and 80 GHz with pulse durations of 710, 1.9, and 2.2 ps, respectively [9]. It should be noted that the amplification without patterning effect in QD SOA can be achieved at much lower pump current densities than in QW SOA [22].

4. Optical synchronization of a QDWELL laser dynamics and its influence on QDWELL laser and SOA performance

We investigated theoretically the optical synchronization of the carrier dynamics in the optically injected QDWELL laser and SOA and its influence on the QDWELL laser and SOA performance [23–27]. We solved numerically the LS equations (14)–(17) for the optically injected QDWELL laser [24–26] and Eqs. (17)–(22) for the QDWELL SOA [23, 27] using the typical values of QDWELL parameters and explicit expressions for the carrier scattering rates S_e^{in}, S_e^{out}, S_h^{in}, S_h^{out} [17, 20]. We used the MATLAB environment for the numerical simulations. We obtained a large number of the simulation results for the QW and QD carrier dynamics, QDWELL laser and SOA output signal eye diagrams, QD electron and hole phase trajectories, constellations for the QDWELL laser output 4-quadrature amplitude (4-QAM) modulation signal, QDWELL SOA gain wavelength and power dependence, QDWELL SOA chirp and extinction ratio (ER) dependence on the detuning between the pump and signal waves. They can be found in Refs. [23–27]. In this section, we summarize the main results and present the figures that demonstrate the influence of the optical injection on the QDWELL laser and QDWELL SOA performance.

It has been shown that the sufficiently strong optical injection leads to the synchronization of the electron and hole dynamics in QD [24]. In the transient turn-on regime, the bias current and the optical output pulses are strongly synchronized in the case of a comparatively high optical injection power of several milliwatts and the moderate bias current density $j = 3j_{th}$, where $j_{th} \approx 6.7A/m^2$ is the threshold current density of the QDWELL laser [24]. In such a case, $\rho_e + \rho_h < 1$, and the term describing the inversion of the two-level system with electrons and holes is negative: $(\rho_e + \rho_h - 1) < 0$. It is seen from Eqs. (14)–(16) that the optically induced transitions in QD are dominant, and the QD electron and hole occupation probabilities ρ_e, ρ_h increase. Their increase is limited by the condition $\rho_e + \rho_h = 1$. Then, the first term in Eqs. (14)–(16) vanishes, and the QD carrier dynamics is determined by the spontaneous transition rate $R_{sp}(\rho_e, \rho_h)$ and essentially different carrier scattering rates S_e^{in}, S_e^{out}, S_h^{in}, S_h^{out} in and out of QD. The photons are mainly absorbed as it is shown in Eq. (14). In the case of a low level of optical injection less than 1 mW, large bias current density $j = 6j_{th}$ and a significant detuning $\Delta\nu_{inj} = 5GHz$ close to the RO frequency of 7 GHz electron occupation probability ρ_e strongly oscillates while the QD hole and QW carrier dynamics remains stable.

The large-signal response of the QDWELL laser is important for digital communication systems. In the case without the optical injection, the QDWELL laser large-signal dynamics is determined by the QW carrier densities w_e, w_h and their lifetime [17]. The sufficiently strong optical injection of the master laser is inserted instead of a large bias current provides a large enough QD carrier density [24]. In such a case, a device heating by the bias current does not occur. The radiative transition rates in QD are determined by the optical injection power and the frequency detuning $\Delta\nu_{inj}$ between the master and slave lasers [24]. The fast stimulated transitions in QD are dominant, and the increase of the optical injection power results in the stronger electron and hole dynamics synchronization. The QW carrier density is varying weakly. In the case of the electrical pseudorandom binary sequence (PRBS) input signal, the modulation frequency may be increased up to 30 GHz due to the enhanced RO frequency [24]. The boundaries of the stable OIL regime are determined by the following condition [28]

$$-K\frac{E_{inj}}{E_0}\sqrt{1+\alpha^2} \leq 2\pi\Delta\nu_{inj} \leq K\frac{E_{inj}}{E_0} \tag{23}$$

where E_0 and E_{inj} are the slave and master laser field amplitudes, respectively. The instability occurs and the QDWELL output signal manifests strong oscillations in the case of the large detuning, low bias current density $j = 2j_{th}$, and low optical injection level [25]. The eye diagrams of the QDWELL under strong optical injection power $P_{inj} > 1$mW start to deteriorate at the repetition frequency of 30 GHz [24, 25].

The directly modulated QDWELL laser can be introduced into an analogous optical link (AOL) in the framework of ultra wideband (UWB) radio-over-fiber (UROOF) technology [26, 29, 30]. The UWB high-speed AOL shown in **Figure 7** consists of electrical/optical (E/O) converter, a standard SMF (SSMF) optical fiber, and optical/electrical (O/E) converter [30]. We solved numerically Eqs. (14)–(19) for the AOL based on the optically injected QDWELL laser and standard SMF (SSMF) [26]. We considered the small UWB signal with the modulation frequency of 60 GHz, the zero detuning $\Delta\nu_{inj} = 0$ between the master laser and the QDWELL slave laser radiation, the 4-quadrature amplitude modulation (4-QAM) format, and

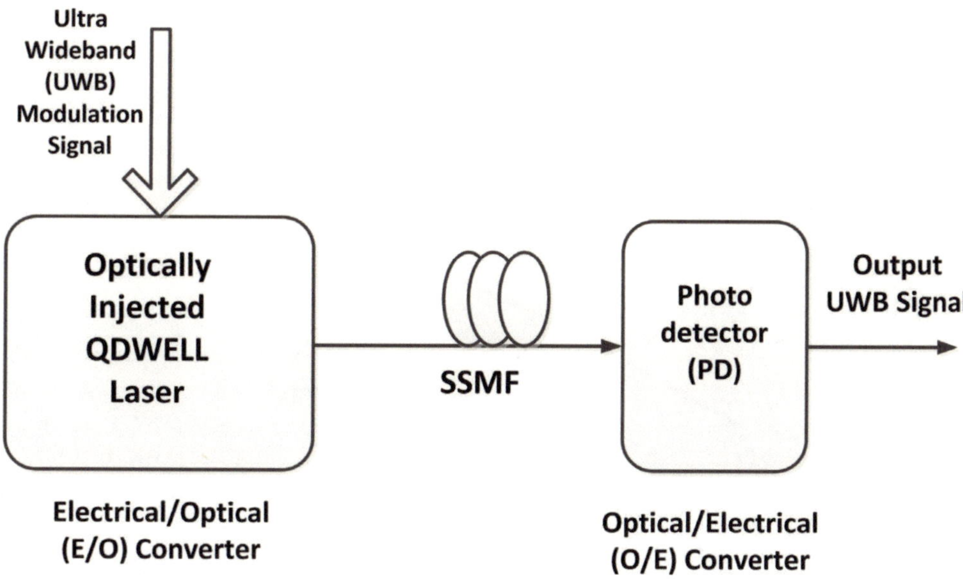

Figure 7. UWB analogous optical link (AOL) with the optically injected QDWELL laser (see **Figure 5**).

128 subcarriers with a 500 MHz bandwidth, and the direct bias current density $j = 2j_{th}$. The dynamics of QD and QW carriers under the strong optical injection of about 1 mW is synchronized, which results in the improvement of the AOL performance.

The calculated constellation of the output 4-QAM modulation signal under the strong optical injection of about 1 mW is shown in **Figure 8**.

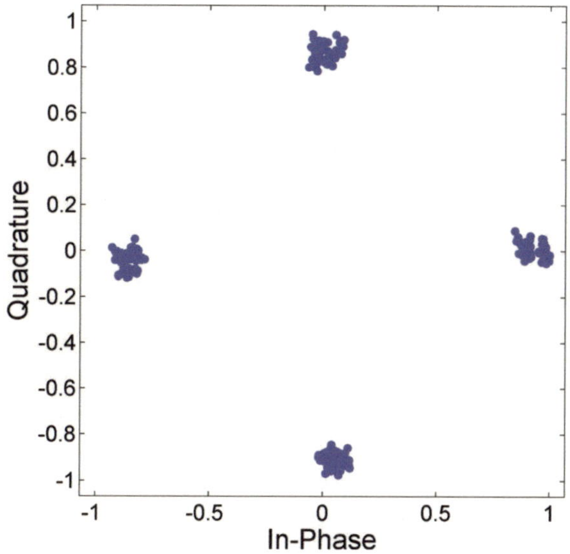

Figure 8. The calculated constellation of the output 4-QAM modulation signal for the modulation frequency 60 GHz, zero detuning, and the optical injection power of 1 mW.

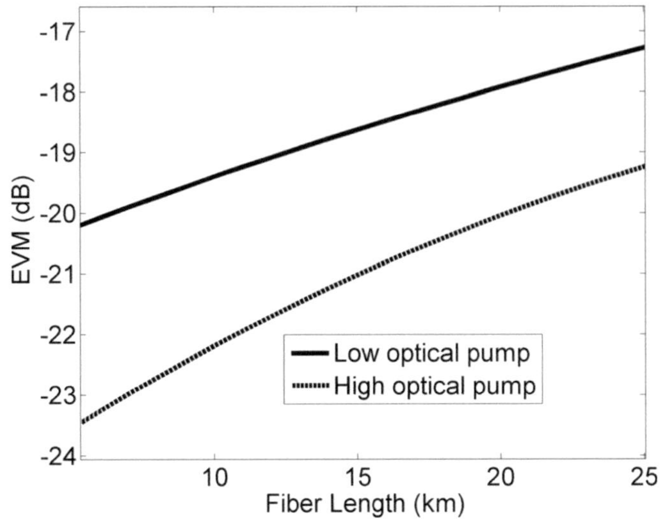

Figure 9. The calculated AOL EVM dependence on the distance for the low optical injection power of 10^{-3} mW (the upper curve) and for the high optical injection power of 1 mW (the lower curve).

The constellation shown in **Figure 8** is practically identical with the constellation of the input signal unlike the case without the optical injection [26]. The AOL error vector magnitude (EVM) dependence on the distance for the optical injection power of 10^{-3} and 1 mW is presented in **Figure 9**.

It is shown in **Figure 9** that in the case of the strong optical injection, the AOL EVM decreases significantly due to the enhancement of the QDWELL laser modulation frequency [26].

The spectrum of the detected UWB signal for the modulation frequency of 60 GHz, zero detuning between the master and the slave lasers, high optical injection power of 1 mW, and the propagation distance of 50 km is shown in **Figure 10** [26].

Consider now the influence of the optical pumping on the QDWELL SOA performance [23, 27]. We investigated the copropagating pumping and signal optical waves characterized by the optical power $P_{p,s}$ and the wavelengths $\lambda_{p,s}$, respectively [23]. Eqs. (17)–(22) have been solved numerically for the pulse regime and the large PRBS signal response [23]. The central pumping and signal wavelengths are $\lambda_p = 1.25$ μm and $\lambda_s = 1.35$ μm, respectively, and the optical pumping power is $P_p = 2, \;\; 5, \;\; 10$ mW [23]. The interaction of the pumping and signal wave in a QDWELL SOA results in the nonlinear optical phenomenon of the cross-gain modulation (XGM) [31]. During the XGM process, the strong modulated pumping wave imposes the modulation on the weak CW signal wave [31]. In such a case, SOA is acting as a wavelength converter (WC) transposing information at one wavelength to the signal at another wavelength [31]. In the case of a QDWELL SOA, the strong pumping wave at the same time synchronizes the QD and QW carrier dynamics. Indeed, when the pumping power $P_p = 1$ μW is small compared to the signal power $P_s = 1$ mW, the electron and hole dynamics in QD is desynchronized, the gain recovery time is determined by the carrier scattering rates

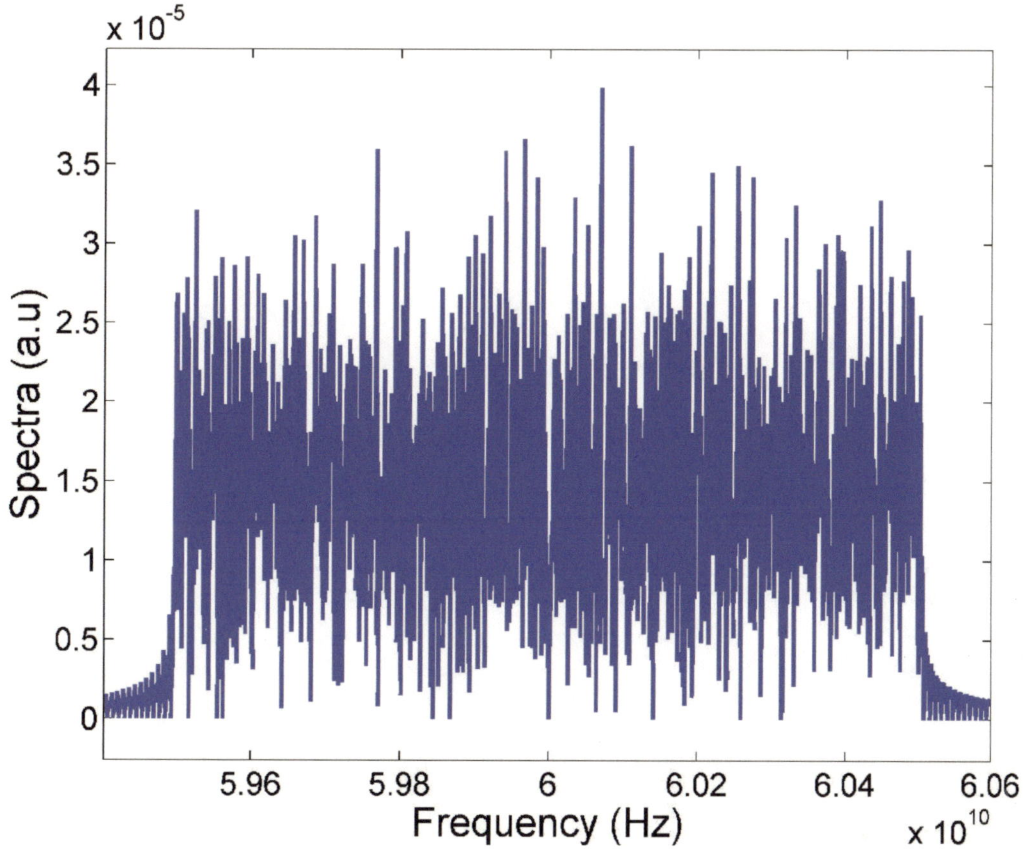

Figure 10. The calculated spectrum of the detected UWB signal after the propagation over the distance of 50 km for the modulation frequency of 60 GHz, zero detuning, and high optical injection power of 1 mW.

S_e^{in}, S_e^{out}, S_h^{in}, S_h^{out}, and the gain recovery process is slow [23]. In the opposite case of the strong pumping power of about 1 mW, the QD electron and hole dynamics is synchronized, the QD carrier occupation probabilities remain practically constant, and the gain recovery time decreases. The oscillation period of the QD carrier occupation probabilities ρ_e, ρ_h is determined by the optically enhanced gain recovery time [23]. The signal wave gain decreases with the increase of the pumping wave power due to XGM and the gain saturation at a large optical power. The bias current density influence on the gain is weak because of the weak connection between QW carrier reservoir and QD during the XGM process [23]. Under the strong optical of 1 mW and the low bias current density of $2j_{th}$, the QDWELL SOA performance improves. The eye diagram of the QDWELL SOA in such a regime is shown in **Figure 11**.

Figure 11 shows that the pattern effect in the eye diagram vanishes up to the repetition frequency of 140 Gb/s. The output signal optical power and chirp time dependence are shown in **Figure 12**. It is shown in **Figure 12** that due to the carrier dynamics synchronization provides the XGM process without the pattern effect, and the chirp reduces to about 5GHz and becomes symmetric [23, 27].

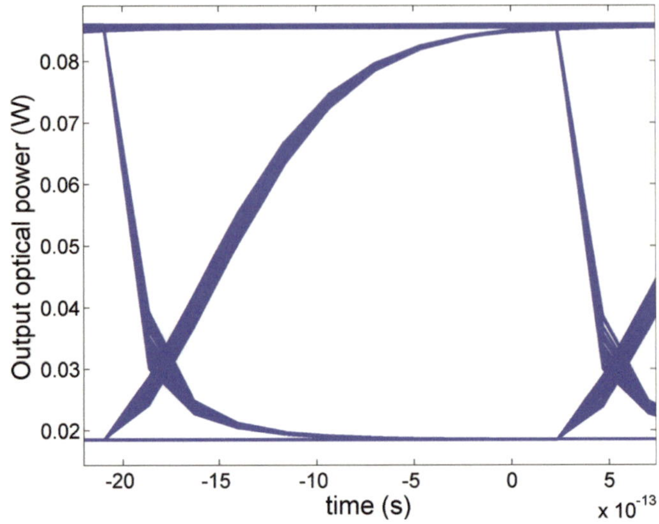

Figure 11. The output signal eye diagram for the signal power of 1 mW, optical pumping power of 2 mW, the bias current density of $2j_{th}$, and modulation bit rate of 140 GB/s.

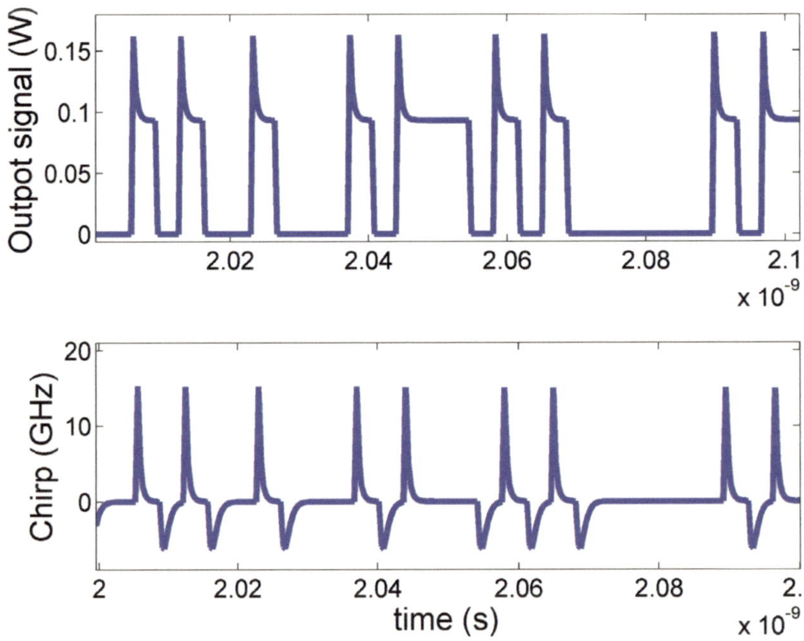

Figure 12. The optical power (the upper box) and the chirp (the lower box) of the output signal.

The strong optical pumping wave of several milliwatts during the XGM process may enhance the QDWELL SOA bandwidth up to 100 nm for a central wavelength of 1350 nm [27]. We investigated the ER dependence on the XGM detuning for the signal wave wavelength

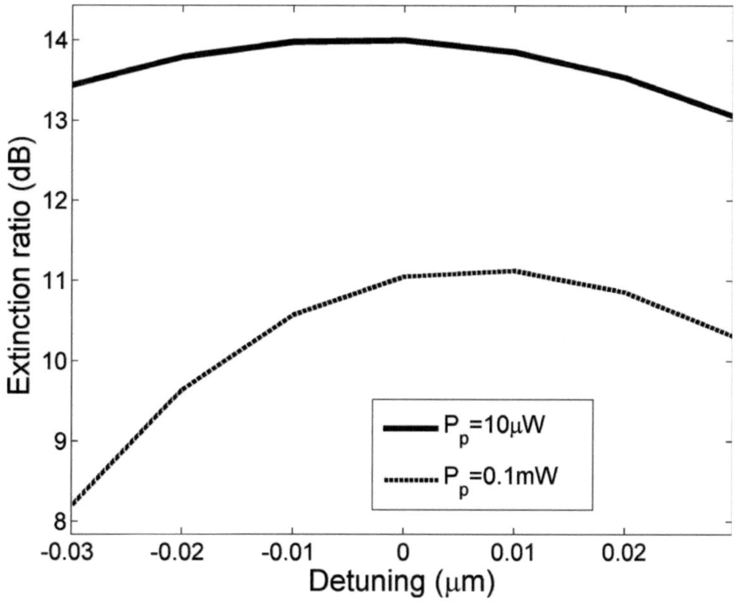

Figure 13. ER dependence on the XGM detuning for the signal wave wavelength $\lambda_s = 1.35$ μm, the signal wave optical power $P_s = 1$ mW, the bias current density of $2j_{th}$, and the CW pumping wave power P_p of 0.1 mW (the lower curve), 0.01 mW (the upper curve).

$\lambda_s = 1.35$μm and optical power $P_s = 1$ mW, and the CW pumping wave power P_p of 0.1 and 0.01 mW [27]. The simulation results are shown in **Figure 13**.

ER reduces with the increase of the CW pumping wave power P_p as it is shown in **Figure 13**.

5. Conclusions

QW and QD semiconductor lasers and SOA are promising candidates for applications in modern optical communication systems due to their comparatively low threshold current, high operation rate, high modulation bandwidth, low chirp, patterning-free operation, and stability with respect to temperature variations. These QW and QD advantages are mainly caused by the one-dimensional (1D) carrier localization in QW and three-dimensional carrier localization in QD. QDWELL lasers are based on the number of QW layers with the embedded QD. In such a case, 2D electron and hole gas also exists in a QW carrier reservoir, i.e., WL. QDWELL lasers are characterized by the lowest possible threshold current and fast QD carrier dynamics due to the 3D localization of electrons and holes in QD. However, the carrier dynamics in QW and QD is desynchronized due to the strong nonlinear carrier scattering rates for the electron and hole transitions in and out of QD. As a result, the modulation frequency of QDWELL lasers is limited by a comparatively low RO frequency of about 7 GHz. The modulation frequency can be enhanced by an increase of the injection current. But the increase of the injection current leads to the QDWELL laser heating and significant performance deterioration.

The modulation frequency and operation rate of semiconductor lasers and SOA can be improved by OIL using the master laser for the optical injection of the slave laser. We solved numerically the system of LS rate equations for QDWELL laser including the optical injection, the explicit expressions for the nonlinear carrier scattering rates, and the inhomogeneous spectral broadening specific for QD. The simulation results show that the carrier dynamics in QW and QD is synchronized by the sufficiently strong optical injection power $P_{inj} > 1$ mW since the fast stimulated transitions become dominant. As a result, the modulation frequency increases up to 30 GHz when the QDWELL output signal eye diagrams start to deteriorate. The high level of the carrier density in QD is provided by the optical injection, and the influence of the bias current is negligible. The strong bias current is not necessary, and the device heating and degradation are avoided. The optical injection process strongly depends on the frequency detuning between the master laser and the QDWELL slave laser. For the low optical injection power and the large detuning of about 10 GHz, the instability occurs accompanied by the strong oscillations of the QDWELL laser output signal. We solved the LS rate equations for AOL based on the optically injected QDWELL laser in the direct modulation regime for the optical injection power of 1 mW, zero detuning, 4-QAM modulation format, and 128 subcarriers with a 500 MHz bandwidth. In such a case, AOL manifests high performance up to the modulation frequency of 60 GHz.

TWA QDWELL LED with an input optical signal operates as a QDWELL SOA. QDWELL SOA is described by the modified system of rate equations including the equations for the pumping and signal wave photon densities. In such a case, the sufficiently strong pumping wave simultaneously plays a role of the optical injection and synchronizes the carrier dynamics in QW and QD. The nonlinear optical processes of XGM and XPM occur in a QDWELL SOA. QDWELL SOA performance is improved, the fast gain recovery takes place, the gain bandwidth is enhanced, ER and chirp decreases.

Author details

Yossef Ben Ezra[1,2] and Boris I. Lembrikov[1]*

*Address all correspondence to: borisle@hit.ac.il

1 Department of Electrical Engineering and Electronics, Holon Institute of Technology (HIT), Holon, Israel

2 MER Cellos, Holon, Israel

References

[1] Agrell E, Karlsson M, Chraplyvy AR, et al. Roadmap of optical communications. Journal of Optics. 2016;**18**(6):1-40. DOI: 10.1088/2040-8978/18/6/063002

[2] Agrawal GP. Fiber-optic Communication Systems. 4th ed. New York: Wiley; 2010. ISBN: 978-0-470-50511-3

[3] Agrawal GP. Nonlinear Fiber Optics. 5th ed. New York: Academic Press; 2013. ISBN: 978-0-12397-023-7

[4] Shieh W, Djordjevic I. Orthogonal Frequency Division Multiplexing for Optical Communications. London: Academic Press Elsevier; 2010. ISBN: 978-0-12-374879-9

[5] Ip E, Pak Tao Lau A, Barros DJF, Kahn JM. Coherent detection in optical fiber systems. Optics Express. 2008;**16**:753-791. ISSN: 1094-4087

[6] Zhao B, Yariv A. Quantum well semiconductor lasers. In: Kapon E, editor. Semiconductor Lasers I. Fundamentals. New York: Academic Press; 1999. pp. 1-121. ISBN: 0-12-397630-8

[7] Harrison P. Quantum Wells, Wires and Dots. Chichester, England: Wiley; 2005. p. 482. ISBN-13: 978-0-470-01079-2

[8] Nagarajan R, Bowers JE. High-Speed lasers. In: Kapon E, editor. Semiconductor Lasers I Fundamentals. New York: Academic Press; 1999. pp. 177-290. ISBN: 0-12-397630-8

[9] Rafailov EU, Cataluna MA, Avrutin EA. Ultrafast Lasers Based on Quantum Dot Structures. Weinheim: Wiley-VCH; 2011. p. 250. ISBN: 978-3-527-40928-0

[10] Lüdge K. Modeling Quantum-Dot-Based devices. In: Lüdge K, editor. Nonlinear Laser Dynamics. Weinheim, Germany: Wiley-VCH; 2012. pp. 1-33. ISBN: 978-0-470-85619-2

[11] Ustinov VM, Zhukov AE, Egorov AY, Maleev NA. Quantum Dot Lasers. Oxford: Oxford University Press; 2003. ISBN: 0 19 852679 2

[12] Ledentsov NN, Bimberg D, Alferov ZI. Progress in epitaxial growth and performance of quantum dot and quantum wire lasers. Journal of Lightwave Technology. 2008;**26**:1540-1555. DOI: 10.1109/JLT.2008.923645

[13] Coleman JJ, Young JD, Garg A. Semiconductor quantum dot lasers: A tutorial. Journal of Lightwave Technology. 2011;**29**:499-5102011. DOI: 10.1109/JLT.2010.2098849

[14] Malić E, Bormann MJP, Hövel P, Kuntz M, Bimberg D, Knorr A, Schöll E. Coulomb damped relaxation oscillations in semiconductor quantum dot lasers. IEEE Journal of Selected Topics in Quantum Electronics. 2007;**13**:1242-1247. DOI: 10.1109/JSTQE.2007.905148

[15] Lüdge K, Bormann MJP, Malić E, Hövel P, Kuntz M, Bimberg D, Knorr A, Schöll E. Turn-on dynamics and modulation response in semiconductor quantum dot lasers. Physical Review B. 2008;**78:**035316-1-035316-11. DOI: 10.103/PhysRevB.78.035316

[16] Lüdge K, Schöll E. Quantum-dot lasers—Desynchronized nonlinear dynamics of electrons and holes. IEEE Journal of Quantum Electronics. 2009;**45**:1396-1403. DOI: 10.1109/JQE.2009.2028159

[17] Lüdge K, Aust R, Fiol G, Stubenrauch M, Arsenijević D, Bimberg D, Schöll E. Large-signal response of semiconductor quantum-dot lasers. IEEE Journal of Quantum Electronics. 2010;**46**:1755-1762. DOI: 10.1109/JQE.2010.2066959

[18] Lüdge K, Schöll E. Nonlinear dynamics of doped semiconductor quantum dot lasers. The European Physics Journal D. 2010;**58**:167-174. DOI: 10.1140/epjd/e2010-00041-8

[19] Lau EK, Wong LJ, Wu MC. Enhanced modulation characteristics of injection-locked lasers: A tutorial. IEEE Journal of Selected Topics in Quantum Electronics. 2009;**15**:618-633. DOI: 10.1109/JSTQE.2009.2014779

[20] Pausch J, Otto C, Tylaite E, Majer N, Schöll E, Lüdge K. Optically injected quantum dot lasers: Impact of nonlinear carrier lifetimes on frequency-locking dynamics. New Journal of Physics. 2012;**14**:1-20. DOI: 10.1088/1367-2630/14/053018

[21] Agrawal GP, Olsson NA. Self-phase modulation and spectral broadening of optical pulses in semiconductor optical amplifiers. IEEE Journal of Quantum Electronics. 1989;**25**:2297-2306. DOI: 0018-9197/89/1100-2297

[22] Uskov AV, Berg TW, Mork J. Theory of pulse-train amplification without patterning effects in quantum-dot semiconductor optical amplifiers. IEEE Journal of Quantum Electronics. 2004;**40:**306-320. DOI: 10.1109/JQE.2003.82032

[23] Ben Ezra Y, Lembrikov BI. Semiconductor optical amplifier based on a quantum Dot-in-a-Well (QDWELL) structure under optical pumping. IEEE Journal of Quantum Electronics. 2014;**50**:340-347. DOI: 10.1109/JQE.20142308393

[24] Ben Ezra Y, Lembrikov BI. Synchronized carrier dynamics in quantum Dot-in-a-Well (QDWELL) laser under an optical injection. IEEE Journal of Selected Topics in Quantum Electronics. 2013;**19**:1-8. DOI: 10.1109/JSTQE.2013.2246770

[25] Ben Ezra Y, Lembrikov BI. Quantum Dot-in-a-Well (QDWELL) laser dynamics under optical injection. Optical and Quantum Electronics. 2014;**46**:1239-1245. DOI: 10.1007/s1 1082-013-9829-3

[26] Ben Ezra Y, Lembrikov BI. UWB system optical link based on a Quantum-Dot-in-a-Well (QDWELL) laser. Optical and Quantum Electronics. 2015;**47**:1527-1533. DOI: 10.1007/s1 1082-015-0154-x

[27] Ben Ezra Y, Lembrikov BI. Investigation of a cross-gain modulation (XGM) in a semiconductor optical amplifier (SOA) based on a quantum Dot-in-a-Well (QDWELL) structure. IET Optoelectronics. 2015;**9**:43-51. DOI: 10.1049/iet-opt.2014.0061

[28] Chang-Hasnain CJ, Zhao X. Ultrahigh-speed laser modulation by injection locking. In: Kaminov IP, Li T, Willner AE, editors. Optical Fiber Telecommunications. V A. Components and Subsystems. London: Elsevier Academic Press; 2008. pp. 145-182. ISBN: 978-0-12-374171-4

[29] Ran M., Ben Ezra Y., Lembrikov B. I. Ultra-wideband radio-over-optical fibre technologies. In: Kraemer R, Katz M., editors. Short-Range Wireless Communications: Emerging Technologies and Applications. Harvard: Wiley; 2009. pp. 271-327. ISBN: 978-0-470-69995-9

[30] Ran M, Ben Ezra Y, Lembrikov BI. High performance analog optical links based on quantum dot devices for UWB signal transmission. In: Lembrikov BI, editor. Ultra Wideband. Croatia: InTechOpen; 2010. pp. 75-96. ISBN: 978-953-307-139-8

[31] Connely MJ. Semiconductor Optical Amplifiers. London: Kluwer; 2002. p. 169. ISBN: 0-7923-7657-9

Chapter 5

Adaptive High Linearity Intensity Modulator for Advanced Microwave Photonic Links

Benjamin Dingel, Nicholas Madamopoulos and
Andru Prescod

Additional information is available at the end of the chapter

http://dx.doi.org/10.5772/intechopen.69262

Abstract

This chapter, first, presents the motivation behind the need for adaptive, highly linear electro-optic modulators and an overview of the different optical linearization approaches of electro-optic modulators. Then, the figures of merits in terms of linearity performance are described and analyzed. Next, the chapter focuses on one excellent linearization approach called interferometric modulator with phase-modulating and cavity-modulating components (IMPACC). Here, we model IMPACC by simulating each of the key building blocks separately before putting them together as IMPACC modulator. This adaptive IMPACC design is compared to typical Mach-Zehnder interferometer (MZI) based modulators, and ring-assisted Mach-Zenhder interferometer (RAMZI) modulators. Theoretical analysis and results show that the IMPACC provides both superior linearity performance and unique adaptive feature that can be used to compensate for manufacturing tolerances, thus, providing extra flexibility in terms of device manufacturability as well as system integration.

Keywords: linear optical intensity modulator, spurious free dynamic range, bandwidth, Mach Zehnder interferometer (MZI) modulator, resonator-assisted Mach Zehnder interferometer (RAMZI), interferometric modulator with phase-modulating and cavity-modulating components (IMPACC), broadband communication

1. Introduction

1.1. Technology trends in microwave photonic links (MPLs)

The current state of microwave photonic links (MPLs) is in the midst of transformation toward more dynamic and flexible MPLs, in response to three major on-going technical shifts, as depicted in **Figure 1**. The first technical shift is an outgrowth of the concept of *software-defined*

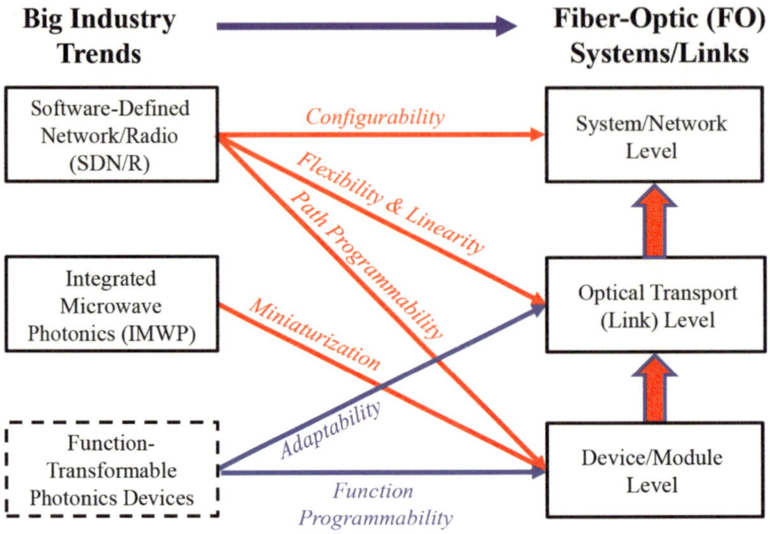

Figure 1. Technology shifts with SDN, integrated microwave photonic (IMWP) and function-transformable photonic devices are generating changes in the architecture, transport and module levels in the traditional fiber-optic links and systems.

networks (SDN) or software-defined radio (SDR) [1–6]. SDN/SDR is going to be investigated, demonstrated, and applied by the microwave photonics community from many different fronts. These include (i) architecture level for future networking needs, (ii) link level to support flexible optical transport layer needs, and even (iii) module level that requires programmable functions that are accessible by direct software control. This shift is driven further by ever expanding applications of MPLs in various "bandwidth-hungry" communication fields like cellular [7] and optical wireless [8], besides the traditional applications of cable television (CATV) [9], subcarrier multiplexing (SCM) optical communication systems [10–12], defense and military [13–16]. The overall goal is toward some form of *software-defined MPLs*.

Figure 2 shows one potential MPL in the context of a bigger configuration defined by SDN—a generic analog optical link, which we will refer to as SD-MPL. It consists of (i) a reconfigurable intensity-modulated direct-detection (IM-DD) link architecture, (ii) control plane, and (iii) network controller. The reconfigurable IM-DD comprises of three modules, namely (1) tunable transmitter, (2) fiber, and (3) receiver. The transmitter module consists of a tunable CW laser source, intensity modulator, RF driver, and RF amplifier, whereas the receiver module could be made up of fiber amplifier and photodetectors for a simple case, and might include DSP module with analog-to-digital converter (ADC) for a more sophisticated/advanced case.

Within the context of SDN, the control plane assembles, enables, and coordinates all pertinent control signals coming from and through the transmitter and receiver modules. Then, the network controller accesses the control plane through OpenFlow interface [5, 17–20] to configure the transmitter and receiver modules. For this generic architecture to function back to traditional, standalone MPLs, the network controller and control plane can be removed and

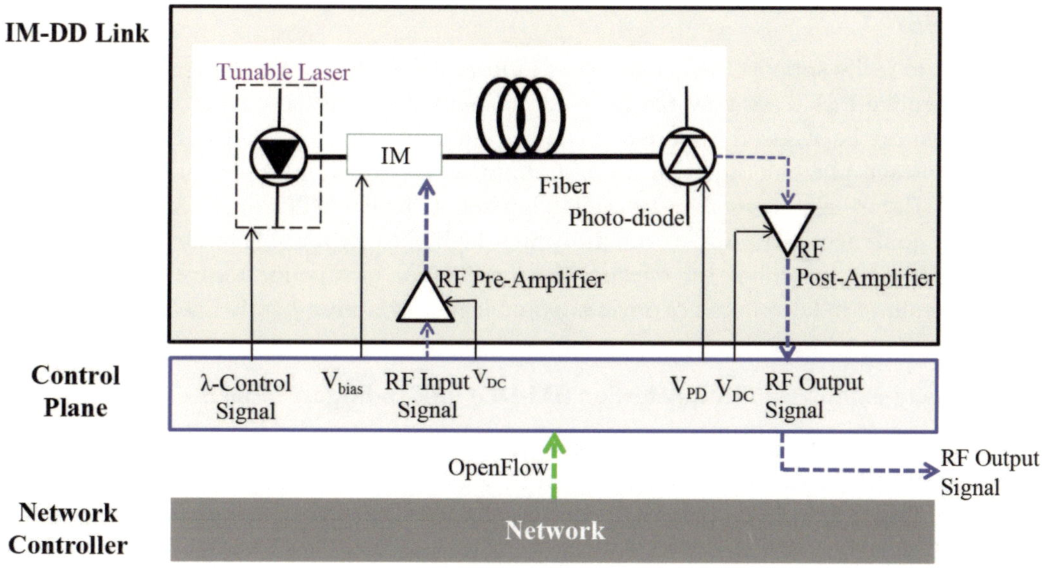

Figure 2. A generic software defined IM-DD-based MPLs in context of a bigger configuration defined by SDN.

the individual control signals are now transferred to the transmitter and receiver modules of the MPLs.

The second technical shift is the rapid growth of the field *integrated microwave photonics (IMWP)* [21–23], which used to be a niche research area some three decades ago. Today, *IMWP* aims to *miniaturize key photonic components, which are essential in the MPLs,* by using the emerging photonic integrated circuits (PICs) [23–26] technology to the level that would be comparable to the performance of state-of-the-art microwave components and systems [21]. Clearly, the rapid developments and advances [22, 23–26] in fabrication technologies re-enforce more advanced IMWP devices and support the prospect of future *software-defined* MPLs.

1.2. Device programmability

The third technical shift is happening at the module level with the development of optical programmable devices and components. These programmable devices and components can have the traditional features of reconfigurability, tunability, or selectability. There are growing number of reports and demonstrations of so-called *microwave photonics processors* [27–33].

Another implementation of these devices is based on the concept of a photonic field-programmable gate array (**FPGA**) or "Photonics FPGA". It is the analogous version of the electronic **FPGA**, which is an integrated circuit designed to be configured by a customer or a designer after manufacturing. "Photonic FPGA" is geared toward arbitrary function generation or function programmability in miniaturized form, which is essential in the full implementation of SDN. This technology is still in the early stages of development, but the growing numbers of demonstrations of on-chip optical processing have been reported and they look promising [27–40].

1.3. Linearity

However, to really support the march toward future IMWP-assisted, software-defined MPLs, it is imperative that these new photonic components not only have inherent *programmable features* that can be accessed by software control, but also *superior linearity characteristics*. Note that microwave photonic links, in its basic form, is essentially a linearization engineering challenge. Regardless of whether we are talking about current MPLs or the future SD-MPLs, the core requirement of the links is still the same—high linearity links! *Superior device's linearity* translates to *high* overall link *performance*. The combination of superior linearity and programmability features in key photonics devices would lead to adaptive, highly linearity SD-MPLs.

1.4. Intensity-modulated direct detection (IM-DD) link and figure of merits

One of the most dominant optical transport schemes in MPLs is the intensity modulated direct detection (IM-DD) because of its simplicity and low cost. IM-DD scheme is shown in the upper portion of **Figure 2**. In its basic form, it consists of a RF data signal, laser source (either in the form of (i) directly modulated laser or (ii) externally modulated laser, using a combination of CW laser transmitter and optical modulator), optical fibers, optical and electrical amplifiers, and photodetector.

For most low-bandwidth (<1 GHz) applications, the directly modulated laser is used as a laser source for economic reason. However, for applications that require higher linearity, larger link gain and wider bandwidth (e.g., 1–60 GHz), the use of an externally modulated laser is the most viable and common approach. It offers the best approach toward achieving higher link linearity compared with directly modulated lasers since CW lasers with very low RIN noise are commercially available. Furthermore, the high output power of a laser can act "as an indirect RF power booster" for the RF input signal to provide a positive gain rather than a negative gain (or equivalent loss) [41]. This important feature adds advantage to the link using an external modulator compared with a directly modulated laser. *From this point forward of this book chapter, when we talk about the links or MPLs, we will specifically mean only externally modulated links.*

The quality of the overall links performance is largely measured by four important figures-of-merit [42], namely: (1) its dynamic range in terms of the spurious-free-dynamic-range (SFDR), (2) link gain (G), (3) noise figure (NF), and (4) bandwidth. The link gain (G) describes the relationship of the RF input signal power to RF output signal power of the IM-DD link, whereas the noise figure (NF) defines the signal-to-noise ratio (SNR) degradation in the link. The SFDR depicts the RF signal power range that can be accommodated by the link, taking into account the effects of noise and nonlinear distortions. It represents the highest signal to noise ratio when the intermodulation signal power is equal to the noise floor. Thus, it combines noise and linearity performance of the link. More detailed technical discussions on these figures-of-merit are found in references [14–16, 43].

The nonlinearities in any part of the links (such as laser, modulator, fiber, photodetector, amplifier, etc.) translate directly to signal noise and distortion. Nonlinear performance at the fiber occurs when the optical powers of the signals are high enough; hence, this is usually mitigated by proper engineering of the optical signal powers to avoid fiber-based nonlinearity. Similarly, detector nonlinearity is mitigated by proper optical power engineering so that the optical power impinging on the photodetector is within the linear response of the photodetector. The nonlinearity of the modulator affects the distortion and dynamic range of fiber optic links. From among these factors, the distortion produced by the external optical modulators has the most negative impact on the overall performance of the link. More specifically, the inherent nonlinear transfer function of the external intensity modulator is the dominant limitation in the performance of MPLs.

1.5. Adaptive linearized intensity modulator

As mentioned earlier, the major source of distortion in IMDD links is the nonlinearity of the intensity modulator. The quest to develop high linearity external modulator with wide bandwidth at low cost has been an on-going challenge for decades.

Besides linearity, other key and important attributes of an external intensity modulator include, (1) low half-wave (V_π) voltage, (2) low insertion loss (*IL*), (3) high bandwidth (*BW*) capability, (4) low-cost, and (5) low complexity design. The low V_π voltage is related to the slope of the transfer function of the modulator. In general, the steeper the slope of the transfer function of the modulator, the smaller the V_π voltage. The immediate impact is an increase of the link gain (G). A modulator with low V_π voltage, as well as low insertion loss, is very attractive since it lowers the necessary input power level of the link and increases gain. Thus, low V_π and low *IL* lead to higher dynamic range, and low noise figure (*NF*), while large bandwidth capability makes it better for high-end applications. Minimal complexity and low cost are interlinked and can open the use of high bandwidth **MPLs** applicable to more commercial applications.

Lastly, any linearized modulator must also have a high tolerance to changes in the RF power level, bias drifts, optical wavelength drifts, etc. This requirement calls for modulator with adaptive capability to provide better maintenance and stable operation of the overall MPLs.

Thus, the focus of this book chapter is two-fold. First, it provides a comprehensive overview of the different linearization approaches that have been proposed and implemented to obtain a highly linear external intensity modulator. Second, it reviews a special type of intensity modulator called interferometric modulator with phase-modulating and cavity-modulating components (IMPACC) that is not only highly linear but also adaptive. These combined features, as far as we know, are not available with any previously reported intensity modulators.

1.6. Objectives and organization

This chapter of the book is organized as follows. First, we give a brief outline of the different linearization technologies used for intensity modulator in Section 2. After a broad overview, we introduce the different optical approaches in linearizing the modulators. Then, we focus on one particular subgrouping known as MZI-based linearized modulator. In Section 3, we present a special type of MZI-based linearized modulator called IMPACC. IMPACC provides both superior linearity and unique adaptive feature. Here, we discuss its basic principle, configuration, and modeling. The comparison and discussion of the original IMPACC and ring-assisted Mach-Zenhder interferometer (RAMZI) modulator are also presented. In Section 4, we provide and analyze IMPACC and RAMZI's respective performances before giving our conclusion in Section 5.

2. Survey of linear intensity modulators

In this section, we discuss the different linear intensity modulators. Before presenting these modulators, we first discuss an important figure-of-merit called spurious free dynamic range (SFDR) which quantifies the linearity of a given link. In Section 2.1, we discuss its derivation using a 2-tone test and graphically describe its significance. In Section 2.2, we provide an overview of the different general techniques to linearize intensity modulators before focusing on one general approach called optical linearization in Section 2.3. Optical linearization has many implementation flavors depending on the particular optical structure used for the modulator. In Section 2.4, we narrow down to a subgroup known as MZI-based modulator because of its popularity and mature technology. Within this subgroup, there are different configurations that have been proposed and implemented, and we spotlight one configuration called IMPACC.

2.1. Measure of linearity: SFDR

A 2-tone frequency test is typically used to evaluate SFDR of the link. **Figure 3(a)** shows a typical IM-DD based MPL consisting of a laser, external modulator RF signal amplifier, optical fiber, photodetector, and RF output amplifier. Note that when the focus of the link is to establish the performance of the external modulator, the common practice is to consider only the intrinsic MPLs. Here, the calculation of dynamic range does not contain the effects of the optical fiber, electrical amplifiers, and optical amplifier, if present.

In **Figure 3(a)**, RF input signal with two closely spaced RF frequencies, f_1 and f_2, of equal-power levels, are simultaneously injected into a nonlinear external optical modulator. The nonlinear interaction between these two signals will create new output frequencies, as shown in the bottom portion of **Figure 3(a)**. Thus, the detected RF output signal will contain two fundamental frequencies, (e.g., f_1 and f_2) together with the second-order IMD output frequencies, (e.g., $f_1 + f_2$ and $f_2 - f_1$), second-order harmonics, (e.g., $2f_1$ and $2f_2$), and third-order IMD outputs (e.g., $2f_1 - f_2$ and $2f_2 - f_1$). Other higher order harmonics and IMD effects (e.g., fourth, fifth, etc.) can also be present, but their effects are generally small and can be filtered out. The second-order SFDR is only important for systems whose bandwidth is more than one octave because the second-order intermodulation products will fall outside the passband of a suboctave system. However, the most important and often most difficult to filter out are the third-order effects which occur very close to the fundamental frequencies within the system bandwidth of interest as shown in **Figure 3(a)**.

Instead of giving the mathematical derivation of SFDR and other relevant parameters or merits, we present a graphical treatment in the discussion of SFDR [16]. For detailed mathematical derivation, we direct readers to relevant references [14–16]. **Figure 3(b)** depicts the output RF

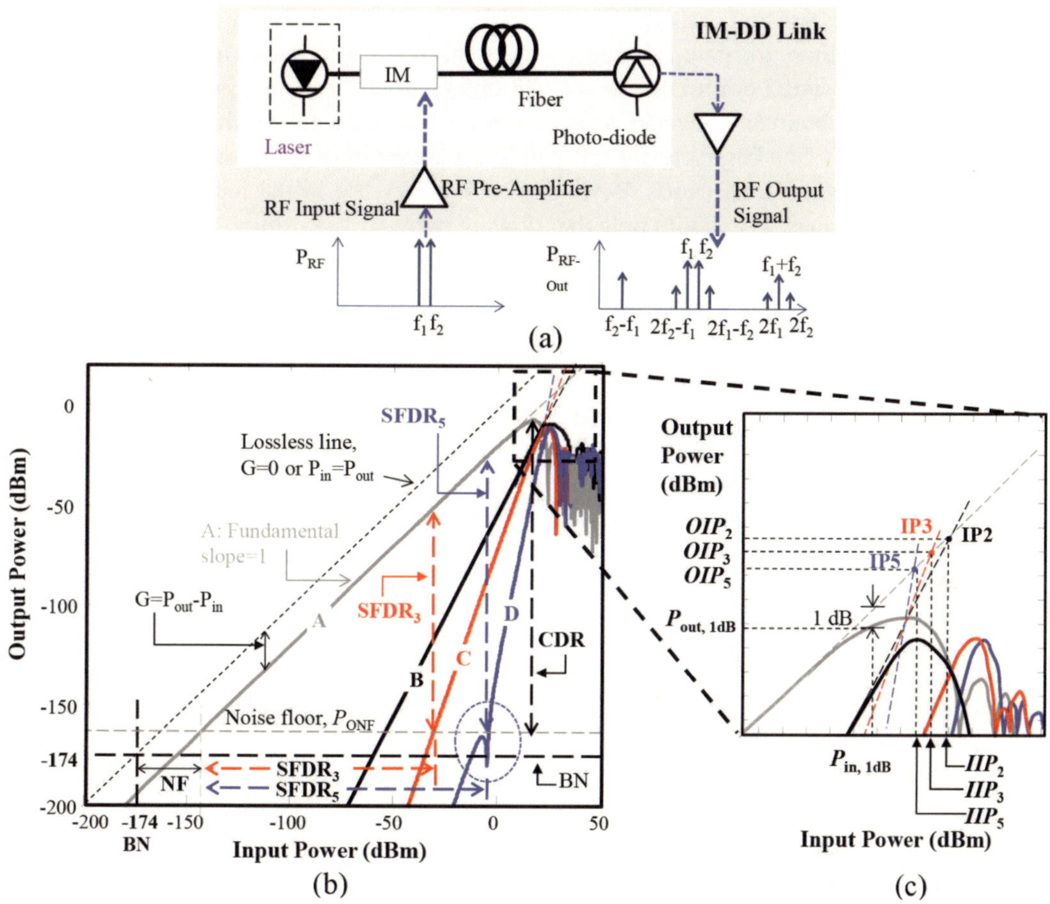

Figure 3. (a) Typical IMDD fiber optic link with input and output RF spectrum; (b) output RF power versus input RF power, showing the important MPL RF parameters/figures of merit, (A) fundamental curve with slope of 1, (B) second-order harmonics and intermodulation terms with slope of 2, (C) third-order intermodulation terms with slope of 3, (D) fifth-order intermodulation terms, with nonlinear slope and not the same slope in the entire frequency range, typical slope at the linear portion is 5 and corresponding SFDR calculations; (c) details of inset of (b) with the definition of the 1-dB compression point, and the *IPn*, with $n = 2, 3, 5$.

signal power versus the RF input power of the different frequency signals generated by the nonlinear interaction within the nonlinear external modulator. This single plot also demonstrates the relationships of the different parameters which we summarize below:

- $\boldsymbol{P_{ONF}}$, ***BN***, $\boldsymbol{P_{ON}}$, and $\boldsymbol{P_{IN}}$: P_{ONF} refers to the output noise floor, which is the minimum discernable signal, whereas ***BN*** is the background noise or thermal noise limit and equal to −174 dBm/Hz at a temperature of 25°C. P_{in} is the RF input signal power and P_{out} is the RF output signal power.

- **Link Gain, G:** The MPL gain, in dB, is defined as the difference of the power at the output of the MPL, P_{out}, to the available power of the input, P_{in} [16]:

$$G = P_{out} - P_{in} \tag{1}$$

We should note here that G is typically referred to as small signal gain, because only in the small-signal regime, the gain is independent of input signal power. That is, for any input power, the associated output power is G dB units higher than the input power. Hence, a straight line is obtained. However, as we continue to increase the modulation power (P_{in}), the straight gain line begins to saturate, until a maximum output power is reached and then starts curving downwards. Beyond this, any increase in the input power leads to lower output power. We should note that G is not necessarily positive. We use it here in its general form, where positive G means increase in the output power, whereas, negative G means reduction in the optical power or loss.

- **Different Signal Slope Lines:** The fundamental curve, third-order IMD curve, and fifth-order IMD curve have a slope of 1, slope of 3, and slope of 5 in the linear region, respectively. Note that oftentimes at the fifth-order IMD, the slope is not necessarily the same throughout the entire frequency band, but it can change from slope 3 to slope 5 [16, 44]. **Figure 3(b)** shows the associated slope lines marked as A (fundamental curve with slope of 1), B (second-order harmonics and intermodulation terms with slope of 2), C (third-order intermodulation terms with slope of 3), and D (fifth-order intermodulation terms, with nonlinear slope and not the same slope in the entire frequency range, typical slope at the linear portion is 5.

- **Noise figure (NF):**
  is defined as the 10 log(F), where F is the noise factor of the MPL and defined as the quotient of the signal-to-noise ratio (SNR) at the input to that at the output, taken for a thermal-noise-limited input at T_s or –174 dBm. Typically this is described by the following equation [16, 43]:

$$NF[dB] \equiv 10\,log(F) = 174 + N_{out}[dBm/Hz] - G[dB], \tag{2}$$

- **Spurious free dynamic range (SFDR):** Graphically, the SFDR is the difference measured from the corresponding point on the fundamental curve to the associated point in the intermodulation distortion (IMD) curve, where IMD intersects or equals the P_{ONF}. Physically, this means that this is the area for which any IMD terms are just below the noise floor and cannot be "seen" by the system. This approach determines the SFDR value, as discussed by Bridges et al. [45]. Mathematically, SFDR is defined as [45]:

$$SFDR = \frac{2}{3}\left(OIP3 - P_{ON} - 10\log_{10}(B)\right) \tag{3}$$

for a system limited by third-order IMD, where OIP3 is the output power at the third-order intercept point and B is the system bandwidth (with units Hz). It is simply two-thirds the difference between the largest distortion-less signal that can be input into the system and the smallest detectable signal in the system. The lower end of this range (the minimum discernable signal) is the sum of the output noise floor (P_{ONF}) and the system bandwidth, B. The measurement bandwidth is assumed to be 1 Hz. In this case, the units of SFDR are dB/Hz$^{2/3}$. However, in the case of more linear systems that are limited by fifth-order IMD, SFDR is mathematically defined as [45]:

$$SFDR = \frac{4}{5}\left(OIP5 - P_{ON} - 10\log_{10}(B)\right) \tag{4}$$

where OIP5 is the output power at the fifth-order intercept point. And, the units of SFDR for a system limited by fifth-order IMD are dB/Hz$^{4/5}$. These respective SFDRs for the third-order and fifth-order are labeled as $SFDR_3$ and $SFDR_5$, respectively in **Figure 3(b)**.

- **1-dB Compression point:** This is the point at which the small signal gain (e.g., straight extended line in **Figure 3(b)**) is 1-dB higher than the actual gain curve. This point signifies the end of the linear regime. A closer look of **1-dB Compression point** is depicted in **Figure 3(c)**.
- **Compression dynamic range (CDR)**: is the difference between the output power at the 1-dB compression point (expressed in dBm) to the smallest such input power (e.g., noise floor).
- ***IP*$_n$–*n*th-order intercept point**: If we extend the linear slope of the fundamental and the *n*th-IMD curve, then these lines intersect at a point. We called this point the *n*th-order intercept point (e.g., IP3 if the IMD is the third-order IMD curve, IP5 if the IMD is the fifth-order IMD curve). The point is defined by the input intercept power (IIP_n) and the output intercept power (OIP_n). These points are depicted in **Figure 3(c)**.

Note that in some new modulator designs, the fifth-order IMD curves are not always straight lines with a constant slope, but have a slight deviation from a straight line just above the noise floor. This is clearly seen in **Figure 3(b)**, in the circled region just below the output system noise level (P_{ONF}—output noise floor), where the slope of an IMD curve increases beyond 5. Since the nonlinearity in the IMD curve is not taken into account in Eqs. (3) and (4), the graphical approach is best suited to be used in calculating the SFDR for the new modulator. In this approach, the SFDR is simply defined as the difference between the fundamental and intermodulation output powers at the input power location where the IMD curve intersects the output noise floor (P_{ONF}).

2.2. Different linearization approaches

Figure 4 summarizes the three general approaches used to linearize a generic external intensity modulator based on the physical mechanism used, namely: (1) electrical, (2) optical, and (3) digital (post detection). The electrical and digital linearization approaches can further be subdivided into smaller categories (as in the case of optical linearization) but we will not review this in much detail here, however more detailed information can be found in [13, 21–22, 41, 46–55].

Briefly speaking, the electrical linearization approach employs different techniques [46–49] such as: (i) RF predistortion, (ii) RF Feed-forward, (iii) combined RF Feed-forward and predistortion, and (iv) others. These techniques generate circuits with either (a) inverse function

(ArcSin) of the modulator to predistort the RF signal before it is injected into the modulator, or (b) split a portion of the RF signal, amplify, reverse its sign, and add it to delayed original RF signal. The overall goal is to cancel or eliminate the original distorted components in the RF signal. These techniques are generally simple and inexpensive but they are limited by the frequency bandwidth of the electrical circuits [51, 53].

On the other hand, a digital linearization approach is based on the application of digital signal processing, DSP [50, 52, 54] techniques to post-detected output signal. The digital approach is very promising because of its flexibility and future growth of DSP. However, its practical usefulness will depend on the progress and development in analog-to-digital converter (ADC) and digital-to-analog converter (DAC) technologies. Unfortunately, ADC is limited by the sampling rate of <10 G/sample at this time [55]. Although there are ADC/DAC technology with >10 G/sample rates, they are usually for instrument purposes [55] and not for field deployment.

In contrast to electrical and digital linearization approaches, optical linearization is free from circuit frequency modulation limitation and does not utilize ADC/DAC technologies. For these reasons, we focus on the optical linearization approach in this chapter.

2.3. Optical linearization

As shown in **Figure 4**, the optical linearization approach can be subdivided further into 5 subgroupings, depending on the key optical structure of the modulator namely; (1) electroabsorption (EA) modulator [56–58], (2) directional coupler (DC) modulator [59–62], (3) all-optical-based approach [63–65], (4) ring resonator modulator [66–72], and (5) Mach-Zenhder interferometer (MZI)-based modulator [73–109]. These different modulators have been implemented and produced different varying degrees of linearity performance. From among these general configurations, the MZI-based modulator is the most well understood, most developed and popular optical structure in the technical community. Hence, we concentrate on the optical linearization approach with a focus on MZI-based linearized modulator.

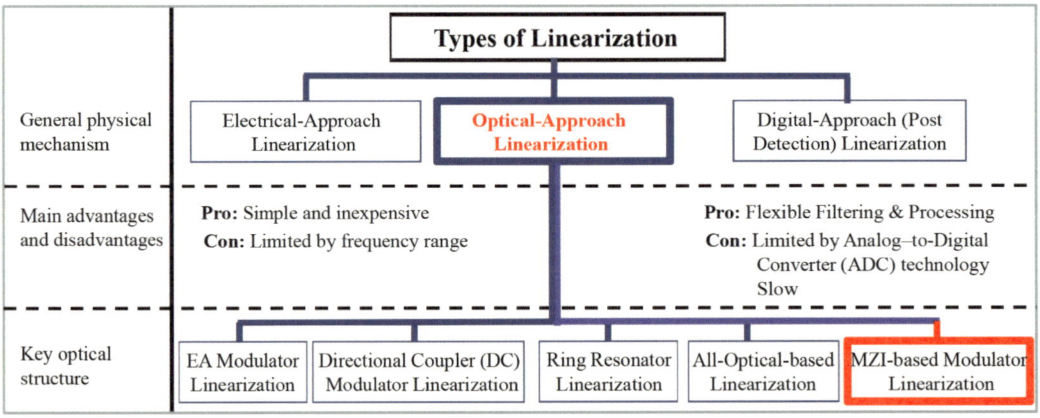

Figure 4. General classification of the different linearity approaches proposed/implemented in fiber-optics links.

http://dx.doi.org/10.5772/intechopen.69262

2.4. MZI-based linearized modulator

Technically speaking, the MZI-based modulator is the most well-studied type of modulator. A typical nonlinearized MZI modulator has a SFDR value of around 90–112 dB-Hz$^{2/3}$. In the last decade, a wide range of linearization approaches has been proposed and implemented to improve the standard MZI modulator to have 120–130 dB-Hz$^{2/3}$ SFDR range using different configurations [74–76, 78–79, 81–84]. Recently, the goal is to push the SFDR further toward >130 dB-Hz$^{4/5}$. Nevertheless, most of them include complex designs and low modulation bandwidth.

Figure 5 shows these different types of MZI-based linearized electro-optic modulators, which can be divided into four groups depending on the actual implementations. The first group can be called dual-signal MZI-based modulator design [74–75, 83–84]. It comprises of one modulator but with two optical signals injected into the device, as depicted in **Figure 5(a)**. The two input signal amplitudes and phases must be properly matched with predetermined values of the RF amplitude phase signal. These two injected optical signals could be implemented using (i) two polarizations [74–75], (ii) two wavelengths [84], or (iii) bidirectional signals [83]. The major advantage of this family of modulators is the relative simplicity and low cost. Its disadvantages are design inflexibility, nonoptimum performance, and tight tolerance requirements.

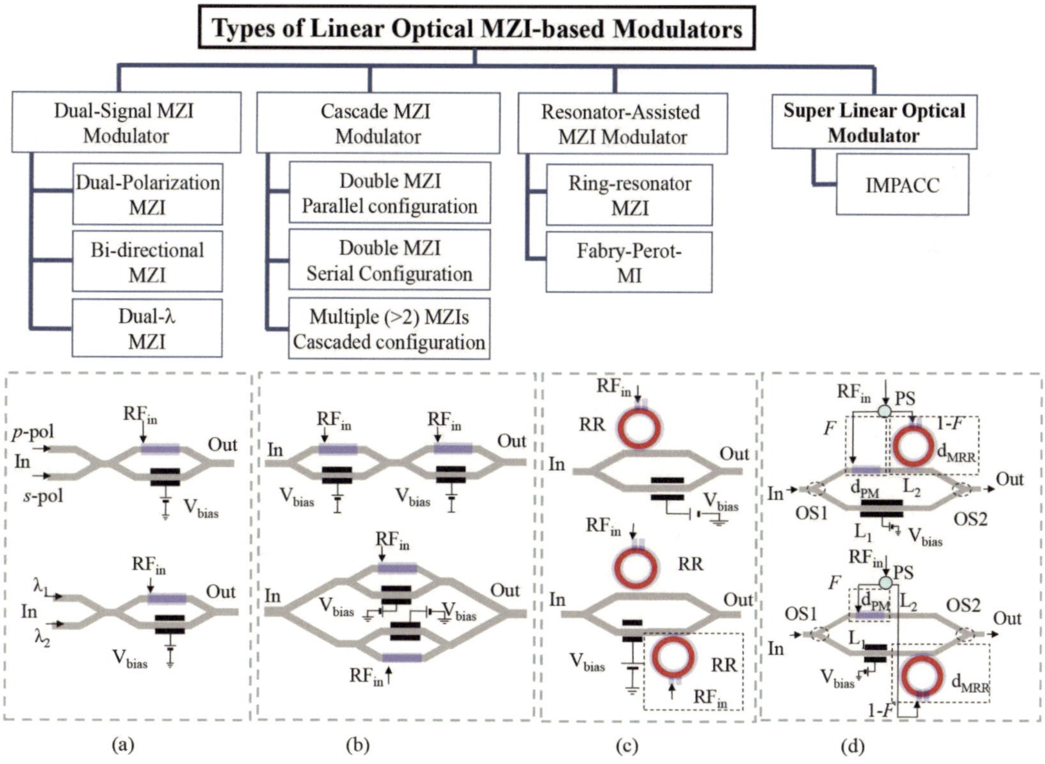

Figure 5. Different types of linear optical modulators based on MZI together with their respective implementation configurations for (a) dual-signal MZI modulator, (b) cascaded modulator, (c) RAMZI, and (d) IMPACC.

The second group can be referred to as cascaded modulator design which consists of two or more standard MZI modulators connected in series [76, 78–79, 80–82] or in parallel [100, 109] arrangements as illustrated in **Figure 5(b)**. It is a generalization of the principle used in the first group. Its major drawbacks are the tight manufacturing tolerance requirements, higher optical loss, higher cost, due to use of multiple modulators, and complicated compensation arrangement.

The third group is generally designated as RAMZI modulators [87–88, 97–99], which has received increased interest in recent years. Some implementations are shown in **Figure 5(c)**. The RAMZI uses ring resonator(s) (RRs) instead of the standard phase modulator (PM) which is coupled in the arm(s) of the MZI. This design gives higher SFDR performance but often at higher manufacturing complexity, limited RF bandwidth range, and stricter transmission coefficient control requirement.

The fourth and last group of optically linearized MZI-based modulators was introduced by our group [89, 93–94, 96, 101–105, 107–108]. This modulator design is a family of modulators which we referred to as **IMPACC**. There are numerous configurations and variations within the family of IMPACC designs [93]. Two of these implementations are shown in **Figure 5(d)**. One of these specific structures of IMPACC is discussed in more details in Section 3 and its performance in Section 4.

3. Design and modeling of IMPACC

This section discusses the basic principle, configuration, and modeling of IMPACC. In Section 3.1, we present an overview of IMPACC configuration. For both educational as well as clarity purposes, we identify the different elements of IMPACC and discuss them separately in Section 3.2. Lastly, we bring all these elements together to come out with IMPACC modulator. While this section gives the basic principle of IMPACC, Section 4 presents its performance.

3.1. Configuration of IMPACC

The configuration of our modulator, known as IMPACC, is shown in **Figure 6**. IMPACC comprises of a MZI configuration, in which one of its arms (e.g., lower) has an active microring resonator (MRR) modulator, whereas the other arm (upper) has an active phase modulator (PM). It uses the standard PM as the phase modulating component and the MRR as the cavity modulating component within a MZI structure [89, 93–94, 96, 101–108]. The exact positions of the PM and MRR relative to one another offer some variations of the implementation, as shown in **Figure 5(d)**.

Interestingly, the IMPACC configuration can be seen as a generalization of both MZI modulator and RAMZI. First, the normal combination of MZI structure and active PM in one of the arm of MZI leads to a typical *single-electrode MZ modulator*. Second, the typical combination of a MZI structure and an active MRR modulator in one of its arms results in a *single-electrode RAMZI modulator* [88, 91–92, 98]. When MRR and PM are present in MZI's arms, then this configuration becomes IMPACC.

http://dx.doi.org/10.5772/intechopen.69262

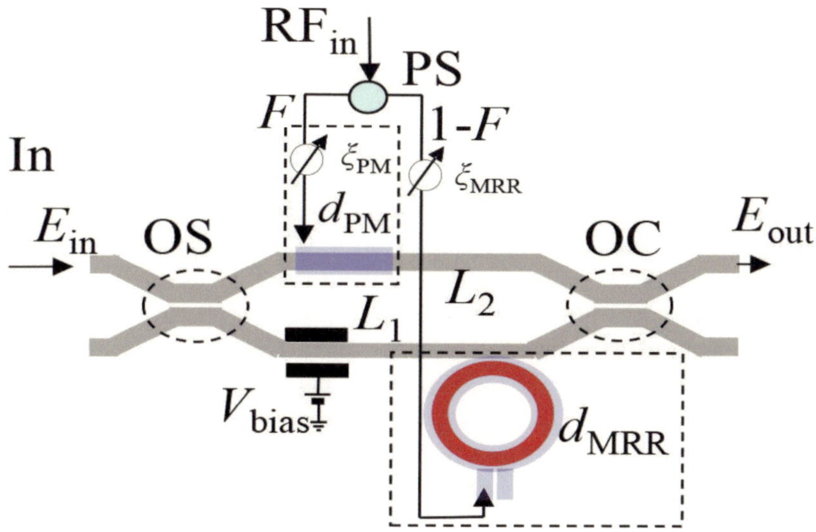

Figure 6. Basic configuration of IMPACC.

Another important aspect of IMPACC is that the RF power signal is split between the MRR and PM by the RF power splitter (PS) with a power split ratio of F:(1−F). This is a critical aspect in the operation of the IMPACC where PM and MRR are driven by the same frequency RF signal, but with different RF power, controlled by the power split ratio (F) using a variable or fixed RF power splitter (PS). Furthermore, the portions of the RF signals directed to MRR and PM electrodes are also delayed by an amount ξ_{MRR} and ξ_{PM}, respectively. These external control parameters from PS are not only instrumental in providing superior linear performance but also have key roles in delivering the unique adaptive mode of IMPACC. This adaptive feature is not available with any known modulators, as far as we know.

The input and output optical splitters (OS and OC) in the MZI configurations are generally assumed to have a (50:50) split so that the optical signal is divided equally into two portions. However, in this chapter of the book, we generalize this condition by taking OS and OC to have arbitrary optical power split ratios namely, P_{OS}:1−P_{OS} and P_{OC}:1−P_{OC}. This condition allows us to investigate the effect of unbalanced power split ratio of OS and OC on the performance of IMPACC.

3.2. Elements of IMPACC

The details of the design of IMPACC have been previously described [77, 89, 93–94, 101–108]. Structurally speaking, IMPACC comprises of four important building blocks namely; (1) Mach Zehnder interferometer (MZI), (2) phase modulator (PM), (3) microring resonator (MRR) modulator, and (4) RF power splitter. In order to appreciate its rich properties, we will first briefly discuss the fundamentals of each of the four important constituents namely; (i) PM, (ii) MZM, (iii) RR, and (iv) RAMZI, before they are combined and operated under certain conditions to create IMPACC.

3.2.1. Phase modulator (PM)

The basic principle of phase modulators is based on the Pockels or linear electro-optic effect [73]. Depending on the material used, the specific axes orientations and the relative alignment of the incoming polarization to the index ellipsoid of the material used, the net effect is the refractive index modulation based on the external applied field. There are two typical configurations, the longitudinal electro-optic modulation and the transverse electro-optic modulation. In the first case, the electric field of the modulating field is parallel to the direction of the optical beam propagation (except near the edges of the electrodes). In the second case, the electric field of the modulating field is perpendicular to the direction of the optical beam propagation. This second case offers refractive index changes induced by the external electric field E that are proportional to the field E. Hence, the electrically induced phase change (or retardation) for the optical beam through the material is proportional to the product EL, where L is the length of the optical beam and electric field interaction and $E=V/d$ is the applied field, with V the applied voltage at the electrodes, which have a separation of d. The net effect is that the ratio L/d can be engineered large enough to provide devices which require low driving voltages.

Referring to **Figure 7**, an incident beam propagating along the y-direction, with polarization along the z-axis (typical case for $LiNbO_3$ based electro-optic devices), will experience an electrically induced phase change, φ_{pm} of [73]:

$$\phi_{PM} = \frac{2\pi}{\lambda} n_e L - \frac{\pi}{\lambda} n_e^3 r_{33} \frac{L}{d} V = \phi_0 - \Delta\phi, \tag{5}$$

where r_{33} is the electro-optic coefficients implicated in this crystal case and orientation [73], λ is the wavelength of the optical beam, n_e is the extraordinary index of refraction seen by the optical beam along this polarization orientation, L is the length of the crystal (or the interaction length of the optical and electric fields), and d is the separation of the electrodes. Note that the total phase change consists of two terms. The first one (e.g., ϕ_0) is the natural propagation phase change, whereas, the second term ($\Delta\phi$) is the electrically induced phase change. Because of the natural birefringence term, a phase compensator is required to adjust the phase until the total phase retardation in the absence of the electric field is an odd multiple of $\pi/2$. The half wave voltage for this case is [73]:

$$V_\pi = \frac{d}{L} \frac{\lambda}{n_e^3 r_{33}} \tag{6}$$

If the applied voltage V is sinusoidal $V = V_m \sin(\omega_m t)$, where ω_m is the modulation frequency and V_m is the applied magnitude of the voltage, then the transmitted beam will be phase modulated which can be expressed as [73]:

$$E_{(y,t)} = A \exp\left[j(\omega t - ky - \phi_0 + \delta \sin \omega_m t)\right] \tag{7}$$

where A is the constant amplitude, and the modulation index δ is given as [73]

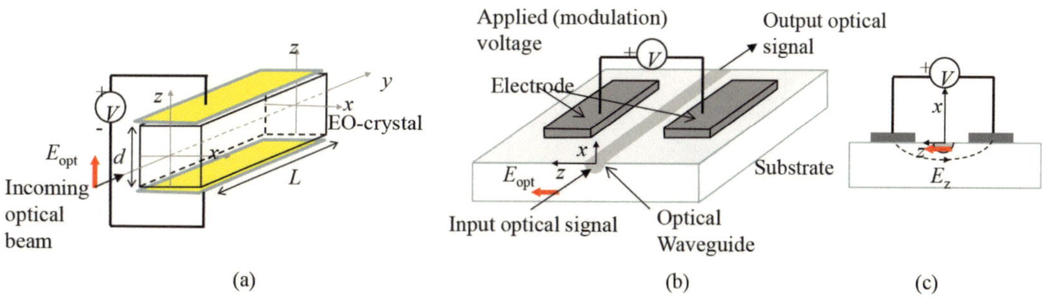

Figure 7. (a) Bulk electro-optic phase modulator, (b) photonic integrated waveguide electro-optic phase modulator in perspective view, (c) waveguide electro-optic phase modulator in cross-section view.

$$\delta = \pi \frac{L}{d} \frac{n_e^3 r_{33}}{\lambda} V_m = \pi \frac{V_m}{V_\pi} \tag{8}$$

We should note here that the phase modulator in a photonic integrated waveguide circuit is slightly different as it has to be modified to accommodate a substrate structure and the electrodes. A typical, but not the only, configuration is shown in **Figure 7(b)** and **(c)**. The optical waveguide is placed at the center of the gap between the two electrodes. In the optical waveguide, the direction of the field of the applied electric field is parallel to the substrate surface along the z-axis. The induced refractive index change Δn_z changes the phase in the way similar to Eq. 5.

3.2.2. Mach-Zehnder modulator (MZM)

A typical external modulator is usually in the form of an integrated optical modulator, which includes a waveguide MZI structure that is fabricated on a slab of lithium niobate ($LiNbO_3$). To ensure that light is channeled through the waveguide, the area is typically doped with impurities in order to increase the index of refraction.

Figure 8 shows a typical MZI modulator configuration where the incident light is split by a Y-branch optical splitrer (OS) into the two arms of the interferometer structure. The light traversing the first arm is modulated by a RF signal. As described in the previous section, the applied electric signal induces a change in the index of refraction of the first waveguide and hence the optical signal experiences a phase shift as it propagates through the waveguide. The magnitude of the phase delay is proportional to the applied voltage and is given by [14]:

$$\Delta\theta_{PM} = \pi \frac{V}{V_\pi}, \tag{9}$$

where V is the voltage applied to the modulator and V_π is the half-wave voltage that when applied, causes a phase shift of π. The second portion of the splitted optical signal propagates into the second arm without RF modulation. Then, the two optical signals are recombined at the output by a Y-branch optical combiner (OC). This recombination can be either constructive

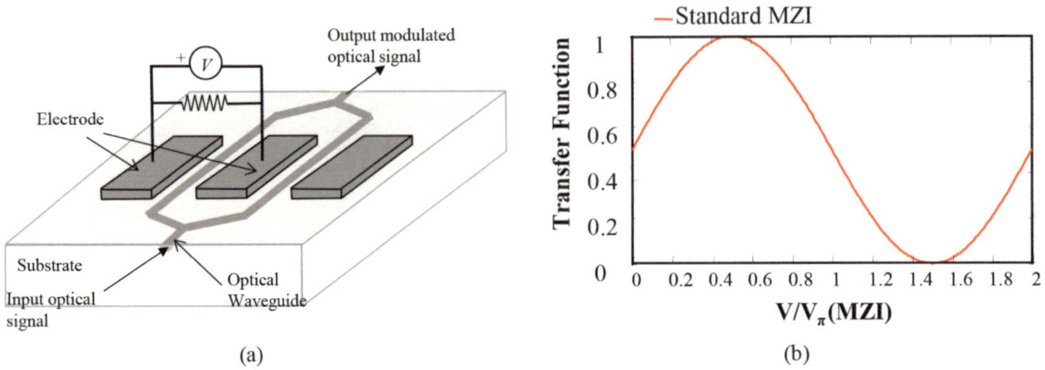

Figure 8. (a) Integrated electro-optic modulation using waveguide Mach-Zehnder Interferometer (MZI) structure, and (b) typical transfer function of a MZI.

or destructive depending on the relative phase difference of the two components of the interfering optical signals. Since the phase of one arm can be controlled by the applied electric signal, the output signal can be modulated. The transfer function of the MZI modulator is given by [73]:

$$T_{\mathrm{MZI}}(V) = \frac{1}{2}\left[1 + \cos\left(\pi\frac{V}{V_\pi} + \theta\right)\right], \tag{10}$$

where θ is the bias phase.

3.2.3. Single microring resonator (MRR)

The basic formulation for the physics of microring resonators (MRR) has been extensively discussed [85–86, 90]. We follow this analysis, in which an MRR has a total circumference length equal to d_{MRR} is coupled to a bus waveguide as shown in **Figure 9(a)**. The MRR round trip phase is given as $\theta_{MRR} = \omega \cdot n_0 \cdot d_{\mathrm{MRR}}/c$, where n_0 is the index of refraction of the material, α is the loss factor of the MRR waveguide, and c is the velocity of the light.

In **Figure 9(a)**, the electric fields E_1, E'_1, E_2, and E'_2 are related by the matrix equation which describes their interactions as [91]:

$$\begin{pmatrix} E_2 \\ E'_2 \end{pmatrix} = \begin{pmatrix} \tau & -i\kappa \\ -i\kappa & \tau \end{pmatrix} \begin{pmatrix} E_1 \\ E'_1 \end{pmatrix} \tag{11}$$

where τ is the transmission coupling coefficient of the coupler between the bus waveguide and the MRR while κ is the associated cross coupling coefficient of the coupler. We assume that only a single mode of the MRR is excited, and the coupler is lossless, so that $\tau^2 + \kappa^2 = 1$.

The output electric field E_2, after *one* round trip, can be expressed as [91]:

$$E_2 = \tau E_1 - i\kappa E'_1 = \tau E_1 - (i\kappa)\alpha e^{-i\theta_{\mathrm{MRR}}} E'_2 = \tau E_1 - (i\kappa)^2 \alpha e^{-i\theta_{\mathrm{MRR}}} E_1 = (\tau - \kappa^2 \alpha e^{-i\theta_{\mathrm{MRR}}}) E_1 \tag{12}$$

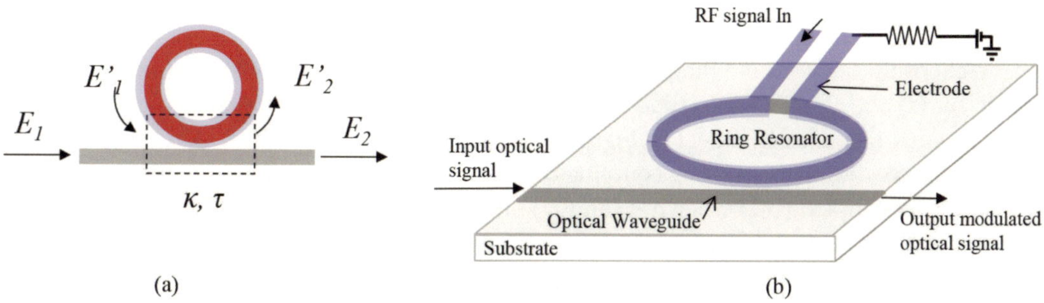

Figure 9. Microring Resonator (MRR) with one bus waveguide, (a) top view showing the field amplitudes at the input and output of the coupling region, (b) perspective view of the MRR with RF electrode.

Here, α is the RR waveguide loss factor ($\alpha = 1$ indicates that the RR is lossless).

After *two* roundtrips, we have [91]:

$$E_2 = (\tau - \kappa^2 \alpha e^{-i\theta_{\text{MRR}}} E_1 - \kappa^2 \tau \alpha^2 e^{-i2\theta_{\text{MRR}}}) E_1 \tag{13}$$

Repeating this circulation around the MRR n times, output electric field E_2 leads to [91]:

$$E_2 = \left[\tau - \kappa^2 \sum_{n=1}^{\infty} \tau^{n-1} \left(\alpha e^{-i\theta_{\text{MRR}}} \right)^n \right] E_1 \tag{14}$$

And finally, after infinite roundtrips, we have got the output electric field E_2 as [91]:

$$E_2 = \frac{\tau - \alpha e^{-i\theta_{\text{MRR}}}}{1 - \tau \alpha e^{-i\theta_{\text{MRR}}}} E_1 \tag{15}$$

Thus, its corresponding normalized transmission transfer function $T_{MRR}(\theta)$ under steady-state condition is given by [91]:

$$T_{\text{MRR}}(\theta) = \left| \frac{E_2}{E_1} \right|^2 = 1 - \frac{(1 - \alpha^2)(1 - \tau^2)}{(1 - \alpha\tau)^2 + 4\alpha\tau \sin^2(\theta_{\text{MRR}}/2)} \tag{16}$$

Accordingly, the corresponding phase response, Θ_{MRR} and group-delay (GD) [95] response of MRR can be derived as [110, 111]:

$$\Theta_{\text{MRR}} = \pi + \theta_{\text{MRR}} + \arctan\left[\frac{\tau \sin(\theta_{\text{MRR}})}{\alpha - \tau \cos(\theta_{\text{MRR}})} \right] + \arctan\left[\frac{\alpha\tau \sin(\theta_{\text{MRR}})}{1 - \alpha\tau \cos(\theta_{\text{MRR}})} \right] \tag{17}$$

$$GD = -\frac{d}{d\omega} [\Theta_{\text{MRR}}] \tag{18}$$

Eqs. (14) and (16) will be handy in the analysis of the new modulator that we discuss in Section 4.

(A).Passive (steady-state) operation

The relationship between the magnitude of the bus waveguide-to-MRR transmission coupling coefficient (τ) and the RR waveguide loss factor (α) defines the three different coupling conditions of MRR [73, 85, 91, 110–112]. **Figure 10(a)** shows the typical MRR while **Figure 10(b)** depicts its three different operating coupling conditions known as (1) critical-coupled (CC) when $\alpha = \tau$, (2) over-coupled (OC) when $\alpha>\tau$, and (3) under-coupled (**UC**) when $\alpha<\tau$. Their respective intensity and phase responses are shown in **Figures 10(b)** and **(c)**, respectively. One notable observation in **Figure 10(b)** is that the intensity output becomes zero at resonance frequency (normalized frequency $\Omega = 0$) under CC condition. In comparison, the respective dip intensity outputs of OC MRR and UC MRR are always higher than zero. The second significant observation is found in **Figure 10(c)** where the phase response of UC MRR is "opposite in direction" compared with phase responses of both OC MRR and CC MRR. This is significant and is used in many group-delay/chromatic dispersion mitigation approaches. Recall that if waveguide lossfactor $\alpha = 1$, then MRR is ideally lossless and becomes an all-pass filter or phase only filter.

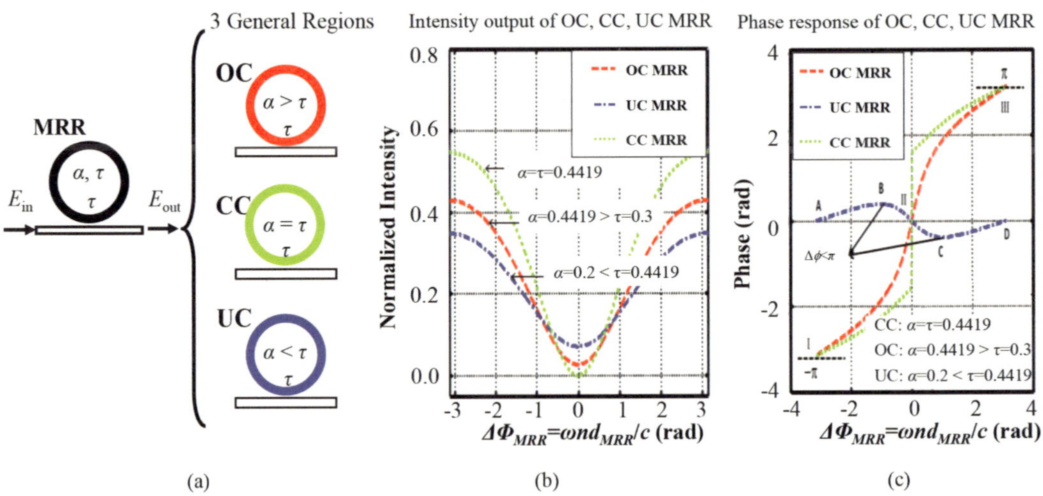

Figure 10. (a)The three different operating coupling conditions of MRR under critical-coupling (CC) at $\alpha = \tau$, over-coupling (OC) at $\alpha > \tau$, and under-coupling (UC) at $\alpha < \tau$ conditions, (b) normalized intensity response for the three operation conditions, and (c) their respective phase responses for the three operation conditions.

(B). Dynamic operation

If a voltage V, is applied to an electro-optic (EO) MRR, a corresponding round-trip phase shift θ_{MRR} induced in the ring resonator is then given as [73]:

$$\theta_{MRR} = \theta_0 + \Delta\beta \cdot d_{MRR}, \tag{19}$$

where $\theta_0 = \omega n\, d_{MRR}/c$ is the static MRR phase response, and

$$\Delta\beta = \frac{\pi n_0^3 r \Gamma V}{\lambda g}. \tag{20}$$

The parameter $\Delta\beta$ is responsible for the dynamic phase change due to applied voltage. Here, r is the electro-optic coefficient of the waveguide, λ is the incident light wavelength, g is the electrode gap and Γ is the overlap integral.

For a sinusoidal voltage, $V=V_0 \sin \omega_m t$ applied to the MRR, then using the same approach that was used to derive Eq. (14), the output amplitude, $E_{\text{out}}(t)$ is given by [77] :

$$E_{out}(t) = \left[\tau - \kappa^2 \sum_{n=1}^{\infty} \tau^{n-1} \alpha^n \exp\left[-i\left(n\theta_{\text{MRR}} + \delta_n \sin(\omega_m t - n\phi)\right)\right]\right] E_{in}(t) \tag{21}$$

where $\phi = \omega_m/FSR$ with ω_m being the modulation frequency and FSR is the free spectral range of MRR, n is the number of round trip around the ring, and δ_n is the modulation index. The modulation index depends on the electrode structure. For a lumped electrode, the modulation index is [91–92]:

$$\delta_n \sin(\omega_m t - n\phi) = \int_0^{nL} \Delta\beta \sin\left[\omega_m\left(t + \frac{n_0}{c} z\right) - n\phi\right] dz = \Delta\beta d_{\text{MRR}} \frac{\sin(n\phi/2)}{\phi/2} \sin(\omega_m t - n\phi) \tag{22}$$

For a traveling wave electrode, the modulation index is given by [91–92]:

$$\begin{aligned} \delta_n \sin(\omega_m t - n\phi) &= \sum_{k=0}^{n-1} \int_0^{nL} \Delta\beta \sin\left[\omega_m\left((t + kt_r) - \frac{\Delta n}{c} z\right) - n\phi\right] dz \\ &= \Delta\beta d_{\text{MRR}} \frac{\sin(\psi/2)}{\psi/2} \frac{\sin(n\phi/2)}{\sin(\phi/2)} \sin\left(\omega_m t - \frac{\psi}{2} - \frac{n+1}{2}\phi\right) \end{aligned} \tag{23}$$

Where $t_r = n_0 L/c$ is the optical round-trip time and $\psi = \omega_m \cdot \Delta n \cdot d_{MRR}/c$ is the velocity matching factor. Here, $\Delta n = n_m - n_0$, where n_m is the microwave effective index and n_0 is the optical index.

3.2.4. RAMZI

Figure 11(a) illustrates a RAMZI configuration where MRR is coupled to a MZI in one of its arms. We assumed that the OS and OC of the MZI have balanced (50:50) power split ratios. Then, the transfer function of RAMZI is given as [66–67, 91–92, 98]:

$$T_{RAMZI}^{SS}(\theta) = \frac{1}{2}\left|\left[\frac{1}{\sqrt{2}}|a(\theta_{\text{MRR}})|e^{-iArg[a(\theta_{\text{MRR}})]} + \frac{1}{\sqrt{2}} e^{-i\phi_{\text{MZI}}}\right]\right|^2, \tag{24}$$

where $\alpha(\theta_{MRR})$ is the complex amplitude transfer function of the MRR under steady-state (SS) condition, which is given by Eq. (14) or (15), the phase response of MRR, $\Theta_{\text{MRR}} = \text{Arg}[\alpha(\theta_{MRR})]$ is given by Eq. (17), and ϕ_{MZI} is the DC phase bias between the two arms of the MZI. In Eq. (24), we assumed that the nominal MZI-arm lengths L_1 and L_2 are equal, except for the length due to DC phase bias. If the MRR is lossless ($\alpha = 1$), $\tau = 2 - \sqrt{3}$ and $\phi_{MZI} = \pi/2$, then

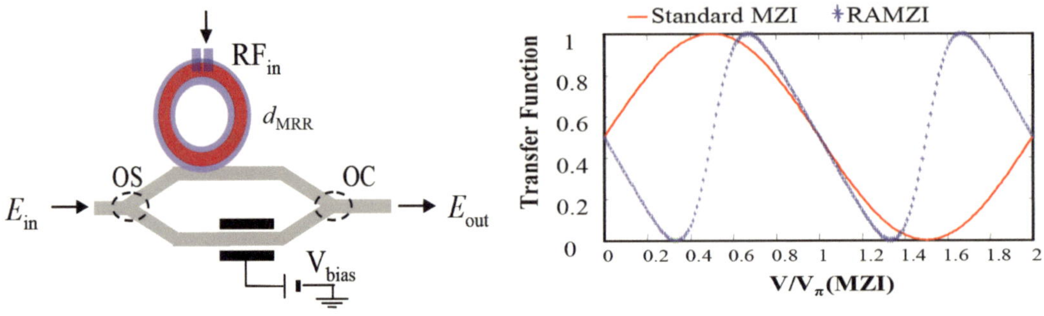

Figure 11. (a)Schematic of RAMZI, and (b) typical transfer function of a RAMZI under steady state condition in comparison with standard MZI.

RAMZI becomes linearized around $\theta = \pi$, which is off-resonance for the MRR [66–67, 91–92, 98]. **Figure 11(b)** shows the typical linearized transfer function of RAMZI in comparison to standard MZI output intensity profile.

On the other hand, under dynamic state (DS) where a modulating sinusoidal RF voltage V is applied to MRR electrode, the corresponding transfer function of RAMZI is given as [91]:

$$T_{RAMZI}^{DS}(\theta) = \frac{1}{2}\left|\frac{1}{\sqrt{2}}|a(\theta_{\text{MRR}})|\exp\left[-i\left(\arg[a(\theta_{\text{MRR}})] + kn_0L_2\right)\right] + \frac{1}{\sqrt{2}}\exp\left[-i(kn_0L_1 + \theta_{\text{bias}})\right]\right|^2, \tag{25}$$

where $a(\theta_{MRR})$ is given by Eq. (21), θ_{bias} is the phase bias of MZI and [91]

$$\theta_{\text{MRR}} = n(kn_0d_{\text{MRR}}) + \Delta\beta d_{\text{MRR}}\frac{\sin(n\phi/2)}{\sin(\phi/2)}\sin\left(\omega_m t - \frac{(n+1)\phi}{2}\right). \tag{26}$$

for a lumped electrode case. The dynamic performance of RAMZI will be shown in Section 4 together with IMPACC performance side-by-side for clear comparison.

3.3. IMPACC

As described briefly in Section 3.1, IMPACC is a generalization of standard MZI modulator and RAMZI. **Figure 12(a)** shows the IMPACC configuration (a) together with its linearized transfer function (b) under its steady-state condition. Here, we assume that both OS and OC have 50:50 optical power split ratios, although later we will relax this restriction and consider arbitrary optical power split ratios namely, P_{OS}:(1−P_{OS}) and P_{OC}:(1−P_{OC}).

Structurally speaking, IMPACC is very much like the RAMZI configuration except for two big differences. First, the passive path-difference, ΔL of the MZI in the RAMZI modulator is now replaced by an active phase modulator (PM) element. Second, the total RF signal driving the electrode in MRR for RAMZI is now split into two portions, by a ratio of F:(1−F), and drives both the electrodes of MRR and PM elements of IMPACC. The two split RF signals are also

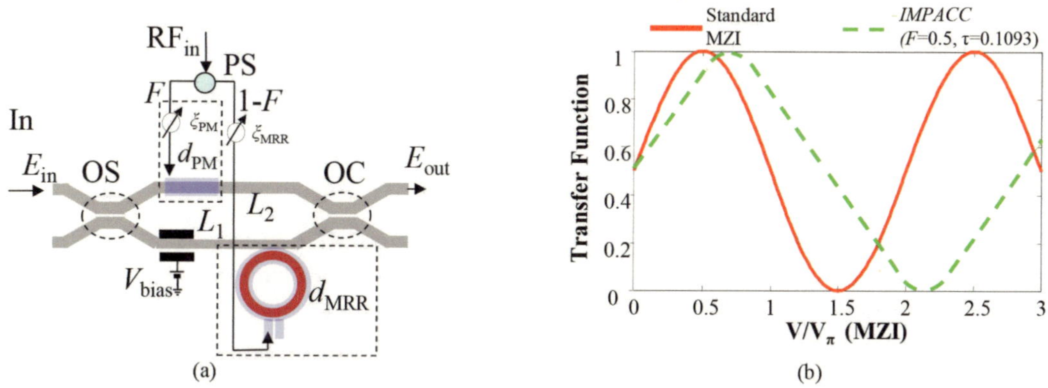

Figure 12. (a) IMPACC configuration, and (b) its transfer function in comparison with standard MZI.

delayed by an amount ξ_{MRR} and ξ_{PM}, respectively. The external control of the RF power split ratio F:(1−F) adds a degree of freedom in the design and serves to drive both the PM and MRR under specific power ratios. The corresponding linearized transfer function of IMPACC under its steady-state condition is shown in **Figure 12(b)**. Here, we assume that both OS and OC have 50:50 optical power split ratios although later in Section 3.3.1, we will relax this restriction by taking optical power split ratios namely, P_{OC}:(1−P_{OC}) to have 50:50 split while OS has arbitrary split of P_{OS}:(1−P_{OS}).

3.3.1. Transfer function

The dynamic transfer function of IMPACC, $T^{DS}{}_{\text{IMPACC}}(\theta)$ is given as [104]:

$$T^{DS}_{\text{IMPACC}}(\theta) = \frac{1}{2}\Big|\sqrt{p_{\text{MRR}}}|a(\theta_{\text{MRR}})|\exp\left[-i(\arg[a(\theta_{\text{MRR}})] + kn_0L_2)\right] + \sqrt{p_{\text{pm}}}\exp\left[-i\left(\varphi_B + kn_0L_1 + \theta_{\text{pm}}\right)\right]\Big|^2 \tag{27}$$

where

$$a(\theta_{\text{MRR}}) = \tau - \kappa^2\sum_{l=1}^{\infty}\tau^{n-1}\alpha^n\exp\left[-i\theta_{\text{MRR}}\right] \tag{28}$$

$$\theta_{\text{MRR}} = n(kn_0d_{\text{mrr}}) + \sqrt{1-F}\Delta\beta d_{\text{MRR}} + \sqrt{1-F}\Delta\beta d_{\text{MRR}}\frac{\sin(n\phi/2)}{\sin(\phi/2)}\sin\left(\omega_{\text{m}}t + \xi_{\text{mrr}} - \frac{(n+1)\phi}{2}\right) \tag{29}$$

and

$$\theta_{\text{pm}} = kn_0d_{\text{pm}} + \sqrt{F}\Delta\beta d_{\text{pm}} + \sqrt{F}\Delta\beta d_{\text{pm}}\sin\left(\omega_{\text{m}}t + \xi_{\text{pm}} - \phi\right) \tag{30}$$

The parameters p_{MRR} and p_{PM} are the input optical power split ratio coefficients of the input optical splitter (OS) for the MRR and the PM arms, respectively. Note that p_{MRR} = 1 − P_{OS}, p_{PM} = P_{OS}, where P_{OS} was defined in Section 3.3. Also, here we assumed that OC has 50:50

power coupling ratio. The parameters θ_{pm} and $\arg[a(\theta_{MRR})]$ are the phase responses of the PM and MRR respectively, while the parameters d_{PM} and d_{MRR} are the lengths of the PM and MRR, respectively with n being the number of times the beam propagates inside the MRR, ω_{m} is the modulation frequency, $\phi = \omega_{\mathrm{m}}/FSR$, φ_B is MZI optical bias phase, ξ_{MRR} and ξ_{PM}, are the RF signal phase biases injected to the MRR and PM electrodes, respectively, and $\Delta\beta$ is given by Eq. (20). We assume negligible microwave loss and no velocity mismatch.

Unlike RAMZI, IMPACC has an external control of either the RF power split ratio or the RF bias. As we will show in the next section, while both of these modulator designs can result in high SFDR performance, IMPACC has an additional and unique property—an adaptive, inherent compensation capability that can be used to maintain its high SFDR under three nonideal parameter conditions namely: (1) increasing MRR waveguide loss factor, (2) unfavorable MRR coupling operating conditions, and (3) unbalanced input power split ratios of either the optical splitter, OS or optical combiner, OC. This adaptive property provides an additional level of design flexibility and can greatly simplify the overall operation of high-performing modulator.

4. Performance of IMPACC

This section provides details of the linearity performance of IMPACC in terms of SFDR, and compares it with RAMZI and standard MZI modulators. Unlike RAMZI and MZI modulators, IMPACC modulator has two different operational modes, namely: (i) the typical basic mode, and (ii) the unique adaptive mode. As far as we know, IMPACC is the only intensity modulator that has an adaptive mode. In Section 4.1, we focus on its basic operating mode, whereas in Section 4.2 we discuss the adaptive mode.

4.1. Basic operating mode of IMPACC

In the basic mode, IMPACC operates under fixed, predetermined parameter conditions aimed to optimize its performance. This mode of operation is typical in all intensity modulators. Here, we investigate the effects of the different parameter variations (from these optimum conditions) on IMPACC's SFDR performance, as well as its bandwidth capability, to assess its performance limits. These parameters include (i) MRR loss factor α, (ii) τ and α that define the CC, UC, and OC conditions, and (iii) unbalanced power split ratio (e.g., P_{os}:$(1-P_{\mathrm{os}})$) of optical splitters (OS). We also compare IMPACC performance with RAMZI and standard MZI modulators.

4.1.1. Effect of ring resonator's loss on SFDR

Any MRR waveguide has an intrinsic loss, (expressed as loss factor, α in Section 3). Here, we assess its effect on both IMPACC and RAMZI's SFDR performance [104]. **Figure 13(a)** shows the respective SFDRs of both IMPACC and RAMZI under two conditions; (a) an ideal lossless MRR waveguide (e.g., $\alpha = 1.0$), and (b) lossy case (e.g., $\alpha = 0.8$). Note that parameter τ is chosen for both IMPACC and RAMZI to get their respective highest SFDR values. We choose to use $\tau = 0.6355$ for

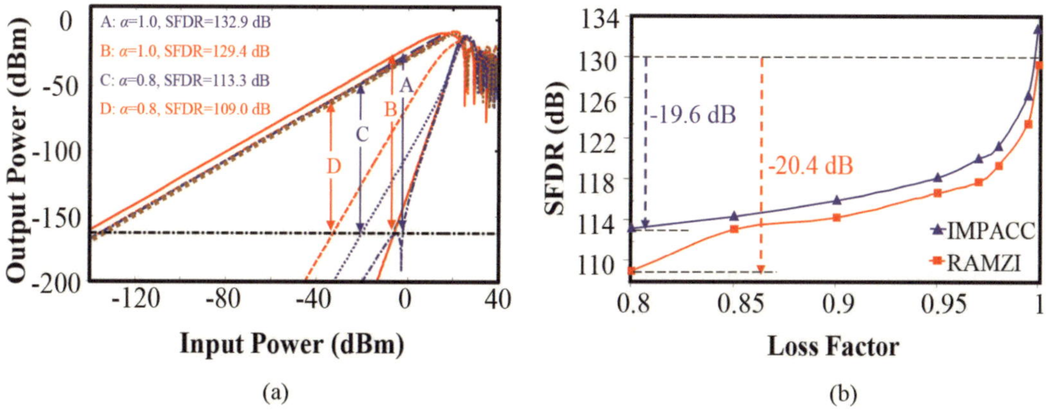

Figure 13. (a) Effect of MRR waveguide loss, α on the SFDRs of IMPACC and RAMZI for ω_m = 1Hz (F, ξ_{PM} are fixed), (A) IMPACC with α = 1, (B) RAMZI with α = 1, (C)IMPACC with α = 0.8, (D) RAMZI with α = 0.8; (b) effect of different values of MRR loss factor on the SFDR values for IMPACC and RAMZI.

IMPACC and the ideal value of τ = 0.2679 for RAMZI. Furthermore, the modulation frequency, ω_m is set to 1 Hz, while the power split ratios of OS and OC are assumed both balanced (50:50). Here, we see that IMPACC has an SFDR value of 132.9 dB under lossless condition (labeled as A) that reduces drastically to 113.3 dB (labeled as C) when loss is introduced. A similar pattern is observed for the RAMZI case, where its SFDR value deteriorates from 129.4 dB for lossless case (labeled as B) to 109.0 dB for lossy case (labeled as D).

Figure 13(b) summarizes the full effect of MRR loss factor α on the SFDR of both IMPACC and RAMZI when the value of α is allowed to vary from 1 to 0.8. We have the following three observations. First, the respective SFDR performances of IMPACC and RAMZI are affected drastically and negatively by the decreasing value of the MRR loss. It is clear from **Figure 13(b)** that the MRR loss value must be minimized within $a < 0.99$ to obtain the higher than 126 dB-Hz SFDR performance. Second, IMPACC always outperforms RAMZI. It exhibits a consistently higher SFDR value (e.g., more than 1 dB) compared than RAMZI for the range $a = 1$ to $a = 0.85$. Third, at the range of $a = 0.85 - 0.80$, the SFDR performance of RAMZI degrades more than three times compared with IMPACC. The comparative SDFR difference between IMPACC and RAMZI increases from 1.2 dB when $a = 0.85 - 4.3$ dB with $a = 0.8$. This implies that IMPACC has a higher parameter tolerance to MRR loss variation especially at higher loss condition.

4.1.2. *Effect of OC, UC, and CC operating conditions on SFDR*

The SFDR performance of any MRR-based devices, including IMPACC and RAMZI, depends highly on the relative ratio between the MRR coupling coefficient strength (τ) and intrinsic loss factor (a) As mentioned in Section 3, there are three MRR regions of operation namely; (i) over-coupling (OC) condition, when $a > \tau$, (ii) critical coupling (CC) when $a = \tau$, and (iii) under-coupling (UC) when $a < \tau$.

Here, we selected the three regions with the following parameters; (i) OC condition (τ = 0.6364, α = 0.7364), (ii) CC condition (τ = 0.6364 and α = 0.6364), and (iii) UC condition (τ = 0.6364 and

$\alpha = 0.5364$). Note again that we choose $\tau = 0.6364$ for IMPACC and the ideal value of $\tau = 0.2679$ for RAMZI. Furthermore in order to simplify the analysis and gain better understanding of the different parameter effects, we assumed an ideal, balanced 50/50 power split ratio for both the input optical splitters (OS) and output optical combiner (OC). The effect on nonideal 50:50 split ratio of OS will be discussed in the next section. Moreover, we assumed that IMPACC and RAMZI are driven at one of its resonant modulation frequency of 23 GHz.

Briefly speaking, the SFDR performance of both IMPACC and RAMZI under OC condition is shown in **Figure 14(a)**. IMPACC has superior linearity with SFDR =130 dB which is far better than RAMZI (e.g., 110 dB). This superior SFDR value is maintained (e.g., 129.7 dB) under CC condition as shown in **Figure 14(b)**. Lastly, the SFDR performance of both IMPACC and RAMZI under UC condition is also shown in **Figure 14(c)**. Under UC condition, both IMPACC and RAMZI are no better than the typical nonlinearized MZI modulator having SFDR value in the range of 110–113 dB. However, we can state that IMPACC's SFDR value (e.g., 113 dB) is still 3 dB higher when compared with RAMZI (e.g., 110 dB). Thus, the takeaway from **Figure 14** is that the huge SFDR advantage of IMPACC to RAMZI exists only when MRR is operated either in OC or CC condition.

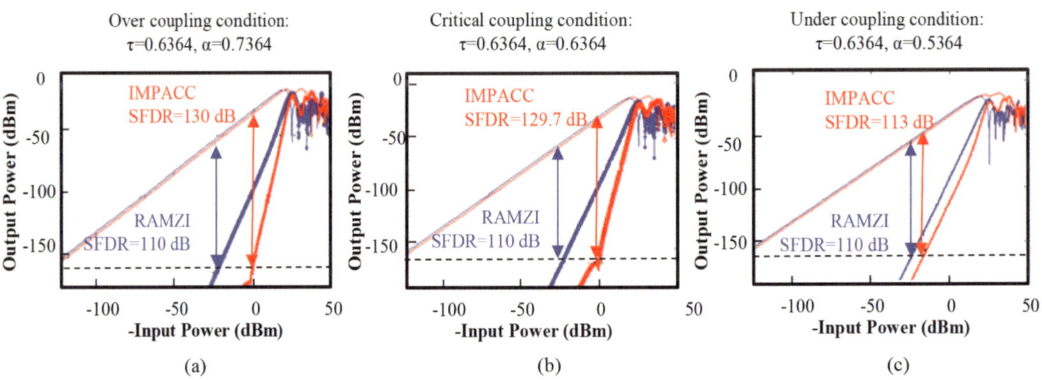

Figure 14. SFDR performance at (a) over-coupling, (b) critical-coupling and (c) under-coupling condition for 23 GHz modulation frequency (analysis assumes a resolution BW of 1 Hz).

4.1.3. Effect of unbalanced optical coupler on SFDR

This section studies the effect on nonideal, unbalanced split ratios of the optical splitter (OS) or optical combiner (OC) on the SFDR performance of IMPACC for three different MRR operating conditions (OC, UC, CC) [110]. To simplify the analysis, we limit our study to unbalanced OS by maintaining OC to have 50:50 power split ratio. Note that we use the same parameters as found in Section 4.1.2 with $\tau = 0.6364$ for IMPACC and the ideal value of $\tau = 0.2679$ for RAMZI.

4.1.3.1. Over-coupled (OC) IMPACC

Figure 15(a) shows the effect of the unbalanced power split ratio (e.g., 55:45) of OS on SFDR performance of IMPACC. The SFDR value drops by 6 dB from 130 dB-Hz under balanced condition to 124 dB-Hz under unbalanced case. **Figure 15(b)** depicts the SFDR performance for various split ratios. It is plotted as a function of the offset from the ideal optical splitter balanced condition (0 deviation corresponds to an ideal power split ratio of 50:50) to a very unbalanced condition (e.g., 0.20 corresponds to power split ratio of 70:30 and −0.20 corresponds to power split ratio of 30:70). One conclusion from **Figure 15(b)** is that in order to maintain a SFDR value of above 124 dB-Hz, we need to ensure that the OS power split ratio deviation should be no greater than ± 0.08 or no more than 58:42 or 42:58 power split ratio. Note that the results of **Figure 15(b)** were obtained without changing any external IMPACC parameters (e.g., RF power split ratio-*F* or the RF phase bias-ξ_{PM}) except the coupling ratio OS.

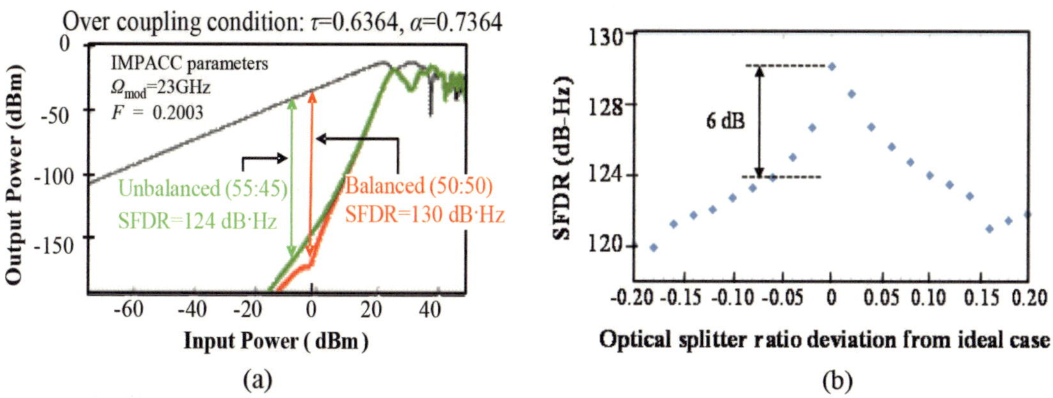

Figure 15. (a) SFDR degradation due to particular unbalanced optical power splitter (55:45) when MRR is operated under over-coupling condition, (b) SFDR performance for different unbalanced optical power split ratios.

4.1.3.2. Critical-coupled (CC) and under-coupled (UC) IMPACC

Figure 16 shows IMPACC's SFDR performance under (a) CC and (b) UC conditions for balanced (i.e., 50:50) and unbalanced (e.g., 55:45) cases. We see that the SFDR value of the OC condition degrades by 11 dB from balanced case of 129 dB-Hz to 118 dB-Hz for unbalanced case. Compared with the OC case (**Figure 15a**) where the SFDR drops only by 6 dB, the CC condition implies higher sensitivity to OS split ratio deviation. On the other hand, **Figure 16(b)** illustrates that OS split ratio deviation has no effect in the balanced or unbalanced cases for UC condition. However, IMPACC has only low SFDR value of 113dB-Hz.

4.1.4. SFDR versus modulation bandwidth response of IMPACC

We know that the modulation bandwidth of all ring resonator based modulators, such as RAMZI and IMPACC is typically limited by the MRR's free-spectral range (FSR) [91, 102, 104–105]. Here, we describe the SFDR-vs-modulation bandwidth capability of both IMPACC and RAMZI, and assess their respective performance limits. We evaluate the SFDR-vs-modulation bandwidth under lossless case (A), and under the three MRR operating conditions (B).

4.1.4.1. MRR lossless case (OC Condition)

Here, we assume that the loss factor *a* in the MRR waveguide, for both the IMPACC and the RAMZI, is lossless or equal to 1. Although, this is not a realistic case, it is done to focus on the

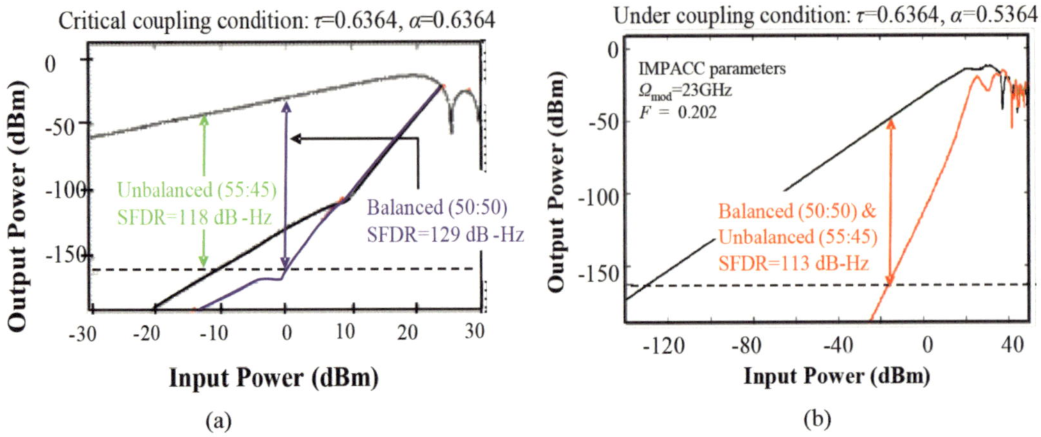

Figure 16. SFDR of IMPACC at over-coupling condition (a), and under-coupling condition (b) for both unbalanced and balanced scenarios (modulation frequency 23 GHz, $\xi_{PM} = 1.3893\pi$, $F = 0.2008$).

bandwidth capability and understand its performance limits. Later in the next subsection, we consider the effect of MRR loss. Here, we use $\tau = 0.208$ for IMPACC, $\tau = 0.268$ (or $\tau = 00.2679$) for RAMZI [90–91, 95, 104], MRR RF phase bias $\xi_{MRR} = 0$, and a PM RF bias control sensitivity ξ_{PM} of 0.2° [105]. The two τ values have been selected such that they provide maximum SFDR value for each case.

First, **Figure 17(a)** shows the SFDR-vs-Modulation frequency of IMPACC (blue line), RAMZI (red line), and MZI modulator (green line) for modulation frequencies up to 75 GHz. IMPACC has an increasingly higher SFDR, (e.g., 3.7 dB to 13.7 dB) compared with RAMZI and standard MZI modulators for these increasing resonant frequencies. It is important to note that the resonance enhancement in IMPACC is accomplished without resorting to a smaller ring size. Smaller ring size not only makes the fabrication more challenging, but would also introduce larger resonator waveguide losses.

Next, we identify two frequency regions of operation: (i) the resonance region at around ~23 GHz (shown in **Figure 17(b)** and ~70GHz (shown in **Figure 17(c)**), and (ii) the non-resonance region at around 56 GHz (shown in **Figure 17(d)**) [105]. The resonance region is defined as the region where the SFDR is greater than that obtained from an ideal MZI, which has a relative flat response as a function of frequency. These resonant regions occur at multiples of the FSR of the modulator (e.g., 23.3 GHz, 69.9 GHz).

As shown in **Figure 17(a)**, IMPACC provides an increasingly higher SFDR, (e.g., 3.7 dB to 13.7 dB) compared with RAMZI and MZI at these resonant frequency regions. **Figures 17(b)** and **(c)** depict the respective IMPACC's SFDR performance when the central RF modulation frequencies are set to 23 GHz and 69.9 GHz, respectively. We note that the modulation linewidth around the central frequency 23 GHz is wider compared with case of central frequency of 69.9 GHz. On the other hand, in the non-resonance region, (defined as the region between the resonance regions), the SFDR is typically less than that obtained from the ideal MZI as shown

http://dx.doi.org/10.5772/intechopen.69262

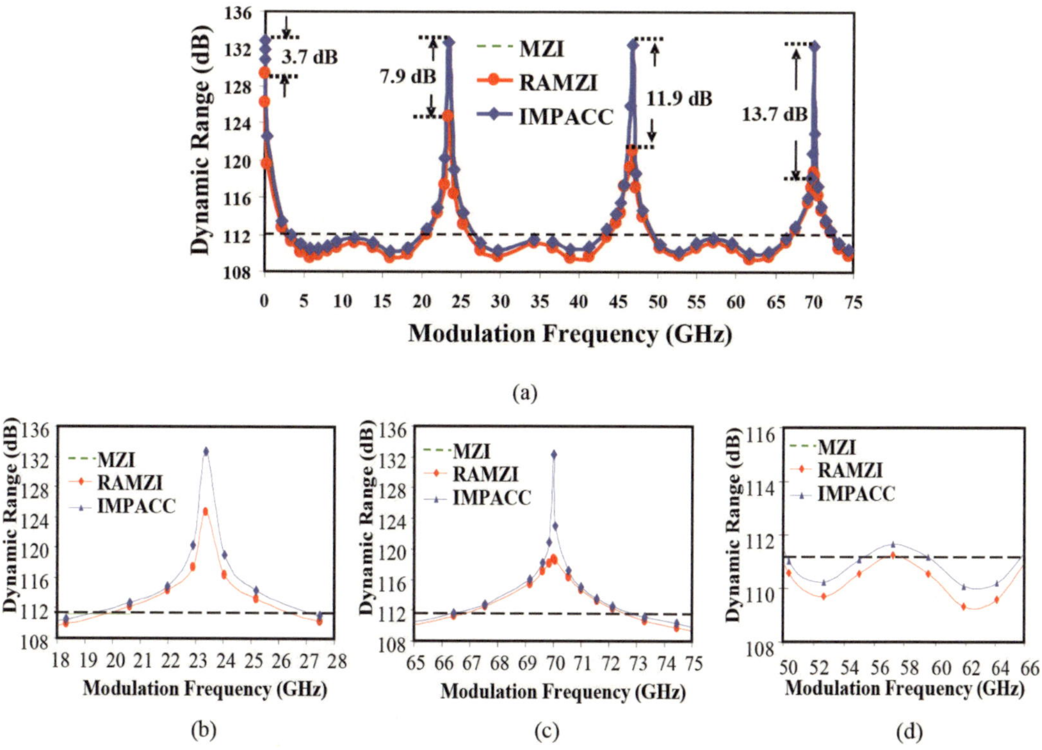

Figure 17. (a) Frequency response for the IMPACC, RAMZI, and MZI, (b) resonant region around a central modulation frequency of 23.3 GHz, (c) resonant region around a central modulation frequency of 70 GHz, (d) non-resonance region around 58 GHz. (analysis assumes a 1 Hz resolution bandwidth).

in **Figure 17(d)**. However, the IMPACC still shows an improved SFDR performance by 0.5 to 2.0 dB compared to RAMZI.

4.1.4.2. MRR lossy cases (OC, CC and UC Conditions)

Figure 18 depicts the SFDR performance of both IMPACC and RAMZI under over-coupled (OC), critical-coupled (CC), and under-coupled (UC) conditions. **Figure 18(a)** shows the OC IMPACC (τ = 0.6364, α = 0.7364) and OC RAMZI (τ = 0.2679, α = 1) as a function of the RF modulation frequency. As we have described earlier, in the regions close to the resonance peak (e.g., ~23 GHz), IMPACC clearly outperforms RAMZI, which operates under ideal condition by 10–20 dB.

The case of CC IMPACC is shown in **Figure 18(b)**. IMPACC also outperforms RAMZI within or in the resonant peak regions. Outside the resonant regions, both IMPACC and RAMZI have nearly the same performance. Lastly, **Figure 18(c)** shows IMPACC performance versus RAMZI in the case of UC condition at the same resonance region as above (e.g., modulation frequency ~23 GHz). In regions close to the resonance peak, the IMPACC still outperforms the ideal over-coupled RAMZI, but now the maximum difference is much smaller (e.g., ~4 dB). We note that

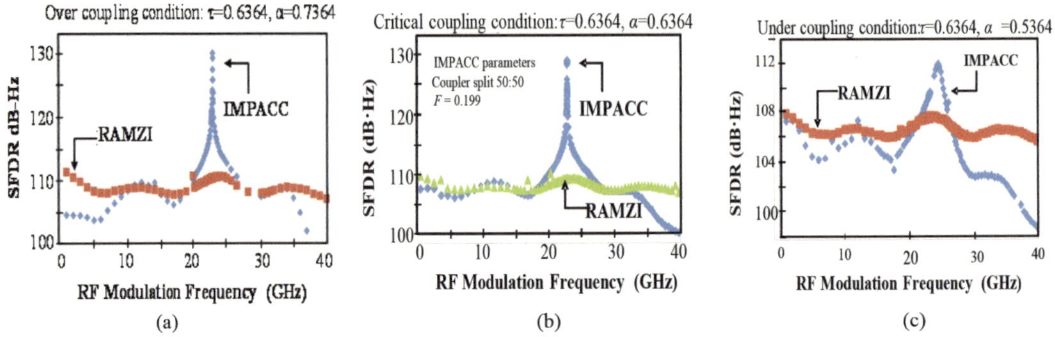

Figure 18. (a) SFDR comparison of IMPACC and RAMZI at (a) over-coupled condition, (b) critical-coupling condition, and (c) under-coupling condition for modulation frequency 23 GHz, $\xi_{PM} = 1.3893\pi$, $F = 0.2008$.

in all these situations, we did not optimize the performance of IMPACC by changing its external parameters (e.g., F or ξ_{PM}).

4.2. Adaptive operating mode of IMPACC

IMPACC's adaptive operating mode is all about using dynamically the two built-in, adjustable, and externally controllable two parameters to maintain IMPACC's 130 dB SFDR performance under unfavorable parameter conditions such as: (i) intrinsic RR loss, (ii) parameter variation of τ value, and (iii) unbalanced OS power split ratio. The unavoidable parameter variations can be due to (a) environmental temperature change, (b) device aging, and (c) fabrication tolerance and errors. The externally controllable parameters used for mitigation purposes are the RF power split ratio F, and the RF phase shift, ξ_{PM}. Here, we set the other RF phase shift, ξ_{MRR} equal to 0, for simplicity. Using again these two external parameters, we show that IMPACC can also extend its SFDR-vs-modulation bandwidth capability beyond that of RAMZI [91, 106–108, 110]. Overall, the adaptive mode of IMPACC allows the link design to be flexible with excellent results.

4.2.1. Extending parameter tolerance to ring resonator loss

The ability to adjust the control parameters F and ξ_{PM} can be exploited to compensate for the negative effect of loss factorα of MRR. In Section 4.1.1, we showed that the SFDR performance of both IMPACC and RAMZI can deteriorate by as much as ~20 dB when the RR loss factor (e.g., α< 1) increases. This effect is reproduced in **Figure 19(a)** for IMPACC and RAMZI with fixed F values. **Figure 19(a)** also shows the SFDR performance for IMPACC with dynamic or adaptive F parameter value. The two cases with fixed F parameter value are similar to the one shown in **Figure 13(b)** and discussed in Section 4.1.1.

The result of this adaptive mode is to maintain the SFDR value to higher than 130 dB by dynamically changing the required F value as a function of the MRR loss factorα as shown in **Figure 19(b)**. It shows the range of required changes in the value of F to achieve the SFDR performance shown in **Figure 19(a)**. This unique ability of IMPACC overcomes the negative effect of the intrinsic insertion loss due to MRR. This feature is not feasible in RAMZI or any

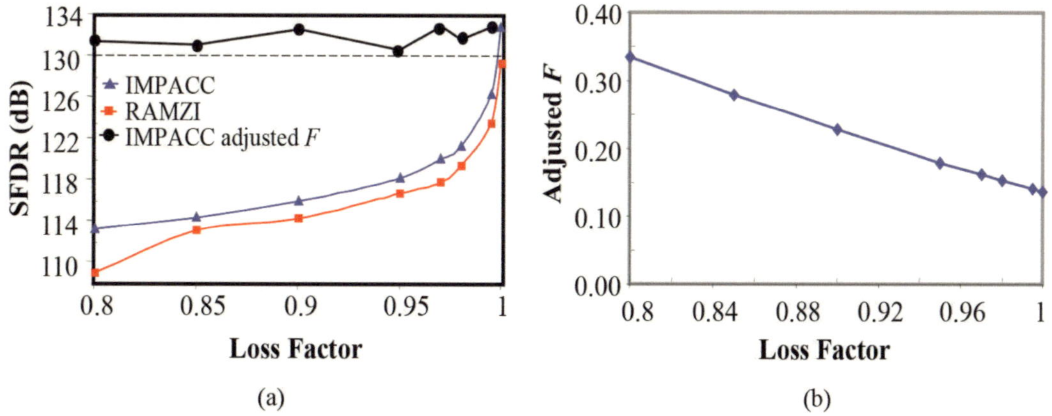

Figure 19. (a) IMPACC compensates for MRR intrinsic losses to obtain SFDR values above 130 dB, (b) RF power split ratio adjustments required to achieve high SFDR (e.g., >130 dB) performance under MRR with exhibited insertion losses.

previously reported linear intensity modulators. It makes the link design and engineering more flexible.

4.2.2. Extending parameter tolerance to unbalanced split ratio of OS

4.2.2.1. Over-coupled (OC) IMPACC and critically-coupled (CC) IMPACC

In this section, we highlight IMPACC's ability to compensate the negative effect due to unbalanced optical splitter performance [110, 113]. **Figure 20(a)** shows the SFDR performance of OC IMPACC (τ = 0.6364, α = 0.7364) for unbalanced OS (55:45; green line), balanced (50:50; uncompensated; red line), and unbalanced (55:45; blue line) but compensated by adjusting the F parameter.

This is emphasized in **Figure 20(a)** where the original drop of 6 dB in the SFDR performance, when the optical splitter is nonideal (55:45), can be compensated to return to SFDR value equal to 129 dB by adjusting the external parameters, F and ξ_{PM} to 0.2003 and 1.307π, respectively.

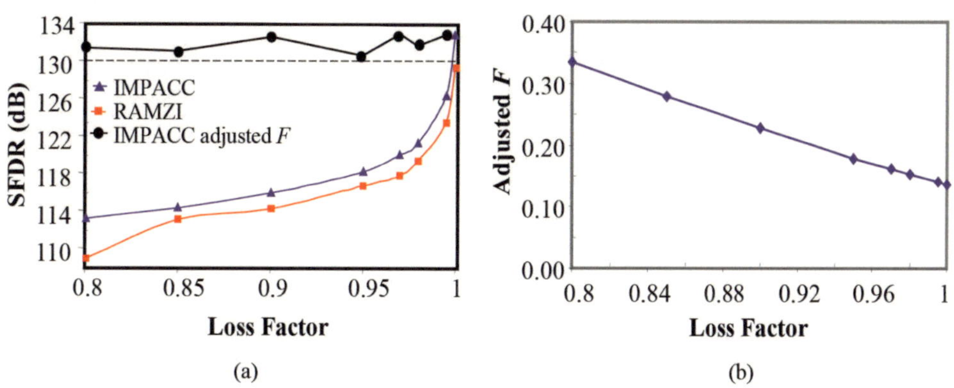

Figure 20. SFDR comparison of (a) over-coupled and (b) critical-coupling condition for RAMZI and IMPACC for both unbalanced and balanced scenarios (modulation frequency 23 GHz, ξ_{PM} = 1.3893π, F = 0.2008).

Hence, clearly, IMPACC's inherent compensation capability is quite effective and able to overcome this limitation. **Figure 20(b)** depicts the SFDR performance of IMPACC at a power balanced for the case of CC condition (i.e., 50:50), unbalanced condition (e.g., 55:45) without compensation and with compensation for the case of CC condition. In the unbalanced condition, the SFDR drops 11 dB from 129 to 118 dB. Nevertheless, IMPACC's inherent compensation capability enables one to increase the SFDR back up to 129 dB for the unbalanced (55:45) case under CC by adjusting the F and ξ_{PM} parameters.

4.2.2.2. Under-coupled (UC) IMPACC

In UC condition, the inherent compensation capability of IMPACC is severely limited and cannot increase the SFDR of IMPACC for the unbalanced case. Adjusting the external parameters,F or ξ_{PM} do not improve the SFDR value. Similar to previous Section 4.1.3, we observed that the IMPACC SFDR performance (e.g., 111 dB) is still higher than that of RAMZI (e.g., 108 dB) but the difference is noticeably smaller compared with the over- and critical-coupled conditions with SFDR of around 129–130 dB.

4.2.3. Extending SFDR versus modulation bandwidth response

As discussed earlier in Section 4.1.3, all ring resonator based modulators (e.g., RAMZI and IMPACC) are limited by the FSR of the ring resonator [88, 91–92, 94,102,104–108]. This is graphically represented by the narrow Lorentzian shape of its SDFR profile as depicted in **Figures 17** and **18**. This limitation is troublesome, especially for wideband RF link applications. Fortunately, the IMPACC has significant advantage compared with RAMZI and other linear modulators that relaxes this built-in RF modulation bandwidth limitation due to MRR.

IMPACC can broaden the SFDR profile at its peak value by dynamically adjusting both F and ξ_{PM} for any given modulation frequency band. **Figure 21** shows this remarkable feature for a frequency of operation from 16 GHz to 20 GHz, where the IMPACC can maintain a SFDR of >130 dB. For the results presented in **Figure 21**, we have used λ = 1550 nm and τ = 0.208, and both the RF phase, ξ_{PM} and power split, F parameters were changed in other to maintain the

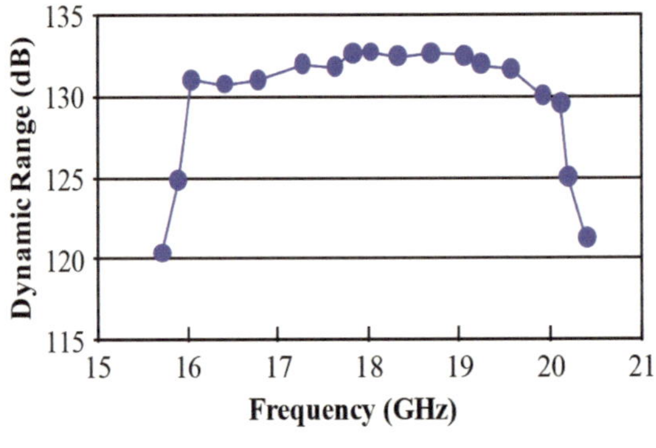

Figure 21. Changing the power split ratio F:(1−F) and RF bias, phase ξ_{PM}, of RF signal can achieve a flat SFDR response over a range of ~20% of around a central frequency (λ = 1550 nm).

high SFDR. We are not aware of any MRR-based modulator designs that can accomplish such a dynamically adjustable wide frequency response using externally controlled parameter.

5. Conclusion

In summary, electro-optic modulators are critical part of optical communications. They will continue to play an important role in fiber optic links. However, an essential requirement is that they have linear response with a high dynamic range of input RF powers of multiple tones. We have shown that IMPACC can operate with highly linear response with peak SFDR value of 132 dB.

Another important requirement in the link design is the inherent versatility of the device or link to compensate for unexpected changes or unavoidable parameter variations due to environmental temperature change, device aging, and fabrication errors. In this regard, IMPACC has a unique place among previously reported linear intensity modulators since it is the only linear modulator design, as far as we know, that has an inherent compensation capability. We demonstrated this inherent feature by compensating various nonideal and often detrimental effects in the modulator from RR waveguide loss and unbalanced optical inputs to various coupling conditions that could lower manufacturing tolerance and degrade linearity performance if not mitigated. The ability to maintain high SFDR (e.g., >130 dB) under these conditions makes IMPACC a viable candidate for many high-bandwidth RF FO-link applications, and well positioned as ideal linear intensity modulator for software-defined MPLs

Acknowledgements

NM would like to thank partial support from the European Committee under program FP7-MC-CIG 333829.

Author details

Benjamin Dingel[1,2]*, Nicholas Madamopoulos[3,4] and Andru Prescod[5]

*Address all correspondence to: bbdingel@stny.rr.com

1 Nasfine Photonics Inc., Painted Post, NY, USA

2 Ateneo Innovation Center, Ateneo De Manila University, Quezon City, Philippines

3 Department of Aeronautical Science, Hellenic Air Force Academy, Dekeleia, Greece

4 Department of Electrical Engineering, The City College of City University of New York, USA

5 ManTech International Corporation, Arlington, VA, USA

References

[1] Elbers J-P, Autenrieth A. From static to software-defined optical networks. In: 16th International Conference on Optical Network Design and Modeling (ONDM), 2012. IEEE; 2012

[2] Gringeri S, Bitar N, Xia TJ. Extending software defined network principles to include optical transport. IEEE Communications Magazine. 2013;**51**(3):32–40

[3] Amaya N, et al. Software defined networking (SDN) over space division multiplexing (SDM) optical networks: Features, benefits and experimental demonstration. Optics Express. 2014;**22**(3):3638–3647

[4] Chen X, Zhang Y. Intelligence on optical transport SDN. International Journal of Computer and Communication Engineering. 2015;**4**(1):5

[5] Kreutz D, et al. Software-defined networking: A comprehensive survey. Proceedings of the IEEE. 2015;**103**(1):14–76

[6] Thyagaturu AS, et al. Software defined optical networks (SDONs): A comprehensive survey. IEEE Communications Surveys & Tutorials. 2016;**18**(4):2738–2786

[7] Lannoo B. Radio-over-fibre for ultra-small 5G cells. In: 17th International Conference on Transparent Optical Networks (ICTON). 2015. pp. 1–4. DOI: 10.1109/ICTON.2015.7193591

[8] Waterhouse R, Novack D. Realizing 5G: Microwave photonics for 5G mobile wireless systems. IEEE Microwave Magazine. 2015;**16**(8):84–92

[9] Way WI. Broadband Hybrid Fiber/Coax Access System Technologies. San Diego, CA: Academic Press; 1998

[10] Al-Raweshidy H, Komaki S, editors. Radio over Fiber Technologies for Mobile Communications Networks. Norwood, MA: Artech House; 2002

[11] Llorente R, Beltrán M. Radio-over-fibre Techniques and Performance. Rijeka, Croatia: INTECH Open Access Publisher; 2010

[12] Liu C, et al. Key microwave-photonics technologies for next-generation cloud-based radio access networks. Journal of Lightwave Technology. 2014;**32**(20):3452–3460

[13] Ackerman E, Cox III C, Riza N, editors. Analog Fiber Optic Links. SPIE Milestone Series. Vol. MS-149; 1998

[14] Chang W, editor. RF Photonics Technologies in Optical Fiber Links. Cambridge, UK: Cambridge University Press; 2002

[15] Vilcot A, Cabon B, Chazelas J, editors. Microwave Photonics from Components to Applications. New York, NY: Kluwer Academic Publishers; 2003

[16] Cox III CH. Analog Optical Links: Theory and Practice. Cambridge, UK: Cambridge University Press; 2004

[17] Ji PN. Software defined optical network. In: 11th International Conference on Optical Communications and Networks (ICOCN), 2012. IEEE; 2012

[18] Channegowda M, et al. Experimental demonstration of an OpenFlow based software-defined optical network employing packet, fixed and flexible DWDM grid technologies on an international multi-domain test bed. Optics Express. 2013;**21**(5):5487–5498

[19] Simeonidou D, Nejabati R, Channegowda MP. Software defined optical networks technology and infrastructure: Enabling software-defined optical network operations. In: Optical Fiber Communication Conference and Exposition and the National Fiber Optic Engineers Conference (OFC/NFOEC); 2013. IEEE; 2013

[20] ONF Solution Brief. OpenFlow-enable Transport SDN. May 2014

[21] Capmany J, Novak D. Microwave photonics combines two worlds. Nature Photonics. 2007;**1**(6):319–330

[22] Marpaung D, et al. Integrated microwave photonics. Laser & Photonics Reviews. 2013;**7**(4):506–538

[23] Capmany J, Munoz P. Integrated microwave photonics for radio access networks. Journal of Lightwave Technology. 2014;**32**(16):2849–2861

[24] Guan B, et al. CMOS compatible reconfigurable silicon photonic lattice filters using cascaded unit cells for RF-photonic processing. IEEE Journal of Selected Topics in Quantum Electronics. 2014;**20**(4):359–368

[25] Iezekiel S, et al. RF engineering meets optoelectronics: Progress in integrated microwave photonics. IEEE Microwave Magazine. 2015;**16**(8):28–45

[26] Bowers JE. Integrated microwave photonics. In: 2015 International Topical Meeting on Microwave Photonics (MWP). IEEE; 2015

[27] Xu X, Bosisio RG, Wu K. Analysis and implementation of six-port software-defined radio receiver platform. IEEE Transactions on Microwave Theory and Techniques. 2006;**54**(7):2937–2943

[28] Ibrahim S, et al. Demonstration of a fast-reconfigurable silicon CMOS optical lattice filter. Optics Express. 2011;**19**(14):13245–13256

[29] Pérez D, Gasulla I, Capmany J. Software-defined reconfigurable microwave photonics processor. Optics Express. 2015;**23**(11):14640–14654

[30] Zhuang L, et al. Programmable photonic signal processor chip for radiofrequency applications. Optica. 2015;**2**(10):854–859

[31] Pérez D, et al. Reconfigurable lattice mesh designs for programmable photonic processors. Optics Express. 2016;**24**(11):12093–12106

[32] Wei W, et al. Software-defined microwave photonic filter with high reconfigurable resolution. Scientific Reports. 2016;**6**

[33] Capmany J, Gasulla I, Pérez D. Microwave photonics: The programmable processor. Nature Photonics. 2016;**10**(1):6–8

[34] Khan MH, et al. Ultrabroad-bandwidth arbitrary radiofrequency waveform generation with a silicon photonic chip-based spectral shaper. Nature Photonics. 2010;**4**(2): 117–122

[35] Burla M, et al. Integrated waveguide Bragg gratings for microwave photonics signal processing. Optics Express. 2013;**21**(21):25120–25147

[36] Pant R, et al. On-chip stimulated Brillouin scattering for microwave signal processing and generation. Laser & Photonics Reviews. 2014;**8**(5):653–666

[37] Shin H, et al. Control of coherent information via on-chip photonic–phononic emitter–receivers. Nature Communications. 2015;**6**. DOI: 10.1038/ncomms7472

[38] Wang J, et al. Reconfigurable radio-frequency arbitrary waveforms synthesized in a silicon photonic chip. Nature Communications 2015;**6**. DOI:10.1038/ncomms6957

[39] Liu W, et al. A fully reconfigurable photonic integrated signal processor. Nature Photonics. 2016;**10**(3):190–195

[40] Wang J, Long Y. A Chip-Scale Microwave Photonic Signal Processing. Rijeka, Croatia: INTECH Open Access Publisher; 2017

[41] Cox CH, et al. Limits on the performance of RF-over-fiber links and their impact on device design. IEEE Transactions on Microwave Theory and Techniques. 2006;**54**(2): 906–920

[42] Iezekiel S, editor. Microwave Photonics: Devices and Applications. Chichester, West Sussex, UK: John Wiley & Sons, Ltd; 2009

[43] Urick Jr VJ, Mckinney JD, Williams KJ. Fundamentals of microwave photonics. In: Chang K, editor. Wiley Series in Microwave and Optical Engineering. Hoboken, New Jersey: John Wiley & Sons, Inc.; 2015

[44] Urick VJ, Diehl JF, Draa MN, McKinney JD, Williams KJ. Wideband analog photonic links: Some performance limits and considerations for multi-octave implementations. In: Nelson RL, Prather DW, Schuetz C, editors. Proceedings of the SPIE, RF and Millimeter-Wave Photonics II; 2012. Vol. 8259:825904

[45] Bridges W, Shaffner J. Distortion in linearized electrooptic modulators. IEEE Transactions on Microwave Theory and Techniques. 1995;**43**(9):2184–2197. DOI: 10.1109/ 22.414563

[46] Ridder RM, Korotky S. Feedforward compensation of integrated optic modulator distortion. In Proceedings of the Technical Digest Optical Fiber Communications Conference, San Francisco, CA; 1990

[47] Chiu Y, et al. Broad-band electronic linearizer for externally modulated analog fiber-optic links. IEEE Photonics Technology Letters. 1999;**11**(1):48–50

[48] Sadhwani R, Basak J, Jalali B. Adaptive electronic linearization of fiber optic links. In: Optical Fiber Communication Conference. Optical Society of America; 2003

[49] Sadhwani R, Jalali B. Adaptive CMOS predistortion linearizer for fiber-optic links. Journal of Lightwave Technology. 2003;**21**(12):3180–3193

[50] Killey RI, et al. Electronic dispersion compensation by signal predistortion using digital processing and a dual-drive Mach-Zehnder modulator. IEEE Photonics Technology Letters. 2005;**17**(3):714–716

[51] Säckinger E. Broadband Circuits for Optical Fiber Communication. John Wiley & Sons; 2005

[52] Moon H, Sedaghat R. FPGA-based adaptive digital predistortion for radio-over-fiber links. Microprocessors and Microsystems. 2006;**30**(3):145–154

[53] Lee T-K, et al. Theoretical analysis and realization of optoelectrical predistortion optical transmitter for the simultaneous suppression of IM3 and IM5 signal. Optics Communications. 2012;**285**(10):2697–2701

[54] Lam D, et al. Digital broadband linearization of optical links. Optics Letters. 2013;**38**(4):446–448

[55] Schmidta C, Kottkea C, Jungnickela V, Freunda R. High-speed digital-to-analog converter concepts. In: Proceedings of the SPIE; 2017, January, Vol. 10130, pp. 101300N-1

[56] Lee G-W, Han S-K. Linear dual electroabsorption modulator for analog optical transmission. Microwave and Optical Technology Letters. 1999;**22**(6):369–373

[57] Betts GE, et al Gain limit in analog links using electroabsorption modulators. IEEE Photonics Technology Letters. 2006;**18**(19):2065–2067

[58] Xie XB, et al. Analysis of linearity of highly saturated electroabsorption modulator link due to photocurrent feedback effect. Optics Express. 2007;**15**(14):8713–8718

[59] Kogelnik H, Schmidt RV. Switched directional couplers with alternating $\Delta\beta$. IEEE Journal of Quantum Electronics. 1976;**12**:396–401

[60] Thaniyavarn S. Modified 1×2 directional coupler waveguide modulator. Electronics Letters. 1986;**22**:941–942

[61] Tavlykaev RF, Ramaswamy RV. Highly linear Y-fed directional coupler modulator with low intermodulation distortion. Journal of Lightwave Technology. 1999;**17**:282–291

[62] Zhang X, et al. Highly linear broadband optical modulator based on electro-optic polymer. IEEE Photonics Journal. 2012;**4**(6):2214–2228

[63] Li J, et al. Optical sideband processing approach for highly linear phase-modulation/direct-detection microwave photonics link. IEEE Photonics Journal. 2014;**6**(5):1–10

[64] Zheng X-P, et al. All-optical signal processing for linearity enhancement of Mach–Zehnder modulators. Chinese Science Bulletin. 2014;**59**(22):2655–2660

[65] Chen J, Zhu D, Pan S. Linearized phase-modulated analog photonic link based on optical carrier band processing. In: 25th Wireless and Optical Communication Conference (WOCC); 2016. IEEE; 2016

[66] Tazawa H, Steier W. Linearity of ring resonator-based electro-optic polymer modulator. Electronics Letters. 2005;**41**(23):1297–1298

[67] Tazawa H, Steier WH. Analysis of ring resonator-based traveling-wave modulators. IEEE Photonics Technology Letters. 2006;**18**(1):211–213

[68] Song M, Zhang L, Beausoleil RG, Willner AE. Nonlinear distortion in a silicon microring-based electro-optic modulator for analog optical links. IEEE Journal of Selected Topics in Quantum Electronics. 2010;**16**(1):185–191

[69] Fegadolli W, Oliveira JEB, Almeida VR. Highly linear electro-optic modulator based on ring resonator. Microwave Optics Technology Letters. 2011;**53**(10):2375–2378

[70] Ayazi A, Baehr-Jones T, Liu Y, Lim AE-J, Hochberg M. Linearity of silicon ring modulators for analog optical links. Optics Express. 2012;**20**(12):13115–13122

[71] Chen L, Chen J, Nagy J, Reano RM. Highly linear ring modulator from hybrid silicon and lithium niobate. Optics Express. 2015;**23**(10):13255–13264

[72] Hosseinzadeh A, Middlebrook CT. Highly linear dual ring resonator modulator for wide bandwidth microwave photonic links. Optics Express. 2016;**24**(24):27268–2727924

[73] Yariv A, Yeh P. Optical Waves in Crystals-Propagation and Control of Laser radiation. New York, NY: John Wiley and Sons; 1984

[74] Johnson LM, Roussell HV. Reduction of intermodulation distortion in interferometric optical modulators. Optics Letters. 1988;**13**(10):928–930

[75] Johnson LM, Roussell HV. Linearization of an interferometric modulator at microwave frequencies by polarization mixing. IEEE Photo Technology Letters. 1990;**2**(11):810–811

[76] Djupsjobacka A. A linearization concept for integrated-optic modulators. Photonics Technology Letters. 1992;**4**(8):869–871

[77] Ho KP, Kahn JM. Optical frequency comb generator using phase modulation in amplified circulating loop. IEEE Photonics Technology Letters. 1993;**5**(6):721–725

[78] Betts GE, O'Donnell FJ. Optical analog link using a linearized modulator. In: IEEE LEOS Annual Meeting, 31 Oct.–3 Nov. 1994; Vol. 2, pp. 278–279

[79] Betts GE. Linearized modulator for suboctave-bandpass optical analog links. IEEE Transactions Microwave Theory Technology. 1994;**42**(12):2642–2649

[80] Sabido DJM, Tabara M, Fong TK, Lu CL, Kazovsky LG. Improving the dynamic range of a coherent AM analog optical link using a cascaded linearized modulator. IEEE Photonics Technology Letters. 1995; **7**(7):813–815

[81] Burns WK. Linearized optical modulator with fifth order correction. IEEE Journal of Lightwave Technology. 1995;**13**(8):1724–1727

[82] Betts GE, O'Donnell FJ. Microwave analog optical links using suboctave linearized modulators. Photonics Technology Letters. 1996;**9**(8):1273–1275

[83] Loayssa A, Alonso M, Benito D, Garde MJ. Linearization of electro-optic modulators at millimeter-wave frequencies. In: Proceedings of the LEOS '99 - 12th Annual Meeting. 8–11 November 1999; Vol. 1, pp. 275–276

[84] Ackerman EI. Broad-band linearization of a Mach-Zehnder electro-optic modulator. IEEE Transactions on Microwave Theory and Techniques. 1999;**47**(12):2271–2279

[85] Yariv A. Universal relations for coupling of optical power between microresonators and dielectric waveguides. Electronics Letters. 2000;**36**:321–322

[86] Okamoto K. Fundamentals of Optical Waveguides. Burlington, MA: Academic Press; 2000

[87] Rengand N, Shpantzer I, Achiam Y, Kaplan A, Greenbalatt A, Harston G, Cho PS. Military Communications Conference. IEEE MILCOM 2003. 13–16 Oct. 2003;2:1208

[88] Xie X, Khurgin J, Kang J, Chow FS. Linearized Mach-Zehnder intensity modulator. IEEE Photonics Technology Letters 2003;**15**(4):531–533

[89] Dingel B. Ultra-linear, broadband optical modulator for high performance analog fiber link system. In: IEEE International Topical Meeting on Microwave Photonics 2004 (MWP'04); 4–6 Oct. 2004. pp. 241–244

[90] Schwelb O. Transmission, group delay, and dispersion in single-ring optical resonators and add/drop filters – A tutorial overview. Journal of Lightwave Technology. 2004;**22**: 1380–1394

[91] Tazawa H. Ring resonator based electro-optic polymer modulators for microwave photonics applications [Ph.D. thesis]. Los Angeles, CA: University Southern California; 2005

[92] Tazawa H, Steier W. Bandwidth of linearized ring resonator assisted Mach-Zehnder modulator. IEEE Photonics Technology Letters. 2005;**17**(9):1851–1853

[93] Dingel B. Linear optical modulator. US Patent 6943931, Sept. 2005

[94] Dingel BB, Madabhushi R, Madamopoulos N. Super-linear optical modulator technologies for optical broadband access network: Development and potential. In: SPIE Proceeding 6012, Optical Transmission Systems and Equipment for WDM Networking IV, October 2005; paper No. 32, Boston

[95] Chen WY, Van V, Herman WN, Ho PT. Periodic microring lattice as a bandstop filter. IEEE Photonics Technology Letters. 2006;**18**(19):2041–2043

[96] Madamopoulos N, Dingel B. Performance analysis of the ultra-linear optical intensity modulator. In: Proceedings of the SPIE 6390, Optical Transmission Systems and Equipment for Networking V, October 2006; paper No. 15, Boston

[97] Van V, Herman W, Ho PT. Linearized microring-loaded mach-zehnder modulator with RF gain. IEEE Journal of Lightwave Technology. 2006;**24**(4):1850–1854

[98] Tazawa H, Kuo YH, Dunayevskiy I, Luo J, Jen A, Fetterman H, Steier W. Ring resonator-based electrooptic polymer traveling-wave modulator. IEEE Journal of Lightwave Technology. 2006;**24**(9):3514–3519

[99] Oliveira JEB, Sakamoto BFR, dos Santos Fegadolli W. Similarities between birefringent gires tournois interferometer and double ring assisted Mach Zehnder. In: European Microwave Conference, Munich; 9–12 Oct 2007. pp. 1330–1333

[100] Zhu G, Liu W, Fetterman HR. A broadband linearized coherent analog fiber-optic link employing dual parallel Mach–Zehnder modulators. IEEE Photonics Technology Letters. 2009:**21**(21):1627–1629

[101] Prescod A, Dingel B, Madamopoulos N. Super-linear modulator with extended bandwidth capability for broadband access applications. In: Proceedings of the SPIE OPTO 2009, Vol. 7234, San Jose, CA; Jan. 24–29, 2009

[102] Prescod A, Dingel BB, Madamopoulos N. Traveling wave optical modulator with high dynamic range (<<130 dB) at large modulation frequencies. In: Proceedings 2009 IEEE Sarnoff Symposium, Session: Optical Communications and Networking, Princeton, NJ; March 30–April 1, 2009. pp. 1–5

[103] Prescod A, Dingel B, Madamopoulos N. An ultra-linear modulator with inherent SFDR compensation capability. In Proceedings of the OSA Integrated Photonics Research, Silicon and Nano Photonics (IPR), July 27, 2010, Monterey, California, USA

[104] Prescod A, Dingel BB, Madamopoulos N, Madabhushi R. Effect of ring resonator waveguide loss on SFDR performance of highly linear optical modulators under suboctave operation. Photonics Technology Letters. 2010;**22**(17):1297–1299

[105] Dingel B, Prescod A, Madamopoulos N. Inherent RF linearized bandwidth broadening capability of an ultra-linear optical modulator. In: Proceedings of the SPIE 7958, Broadband Access Communication Technologies IV; January 2011; paper 7958–3, San Francisco, CA

[106] Dingel B, Madamopoulos N, Prescod A, Madabhushi R. Analytical model, analysis and parameter optimization of a super linear electro-optic modulator (SFDR > 130 dB). Optics Communications. 2011;**284**(24):5578–5587

[107] Madamopoulos N, Dingel B, Prescod A. Interferometric modulator with phase-modulating and cavity-modulating components (IMPACC) for high linearity microwave applications. In Proceedings of the ICPS 2013, The International Conference on Photonics Solutions, Pattaya City, Thailand, May 26–28, 2013 (Co-sponsored by OSA, SPIE, IEEE)

[108] Madamopoulos N, Dingel BB, Prescod A. Review and challenges of high frequency cavity-assisted electro-optic modulators. In: Wireless Optical Communication Conference, WOCC 2014, Newark, NJ, USA; 2014

[109] Jiang W, et al. A linearization analog photonic link with high third-order intermodulation distortion suppression based on dual-parallel Mach–Zehnder modulator. IEEE Photonics Journal. 2015;**7**(3):1–8

[110] Dingel B, Madamopoulos N, Prescod A, Madabhushi R. Power balancing effect on the performance of IMPACC modulator under critical coupling (CC), over coupling (OC), and under coupling (UC) conditions at high frequency. Proceedings of the SPIE Broadband Access Communication Technologies. 2012;**VI**:828208

[111] Ye B, Dingel BB, Cui W. Minimalist-design, high-functionality, micro-ring resonator-based optical filter with narrow linewidth and low group delay using Looped Back Over- and Under-coupled Resonator (LOBOUR). Proceedings of the SPIE. Broadband Access Communication Technologies VII. 2013;8645:86450T

[112] Dingel B, Ye B, Cui W, Madamopoulos N. High-performance, minimalist-design, microring resonator-based optical filter with reduced low group delay and simplified center-wavelength control. International Conference on Photonics Solutions 2013. Thailand; 2013

[113] Dingel B, Prescod A, Madamopoulos N, Madabhushi R. Performance of ring resonator-based linear optical modulator (IMPACC) under critical coupling (CC), over coupling (OC), and under coupling (UC) conditions. In: IEEE Photonics Annual Society Meeting; 2011

Chapter 6

Single-Photon Avalanche Diodes in CMOS Technologies for Optical Communications

Edward M.D. Fisher

Additional information is available at the end of the chapter

http://dx.doi.org/10.5772/intechopen.68935

Abstract

As optical communications may soon supplement Wi-Fi technologies, a concept known as visible light communications (VLC), low-cost receivers must provide extreme sensitivity to alleviate attenuation factors and overall power usage within communications link budgets. We present circuits with an advantage over conventional optical receivers, in that gain can be applied within the photodiode thus reducing the need for amplification circuits. To achieve this, single-photon avalanche diodes (SPADs) can be implemented in complementary metal-oxide-semiconductor (CMOS) technologies and have already been investigated in several topologies for VLC. The digital nature of SPADs removes the design effort used for low-noise, high-gain but high-bandwidth analogue circuits. We therefore present one of these circuit topologies, along with some common design and performance metrics. SPAD receivers are however not yet mature prompting research to take low-level parameters up to the communications level.

Keywords: single-photon, avalanche, diodes, visible, light, communications, receivers

1. Introduction

Optical technologies are routinely used in inter-continental communications, where both data rates and transmission lengths are high, or in fast networking applications such as data centres [1, 2]. At the opposite end of the distance scale, interconnects in microelectronics are moving towards photonics. This is due to the bandwidth restrictions of metals on silicon ICs. By using free-space links for personal computing, optical communication may offer an alternative to wireless standards, i.e. visible-light communications (VLC) applications [3, 4]. The radio spectrum is becoming increasingly crowded, with the UK's frequency allocation tables

(UKFAT) showing severe restrictions on wireless bandwidth, especially as 4G mobile internet is expanded towards 5G. Ultimately, the radio spectrum is limited at the high-GHz and low-THz bands by atmospheric attenuation, principally the molecular resonances of water and oxygen. There can be no doubt that as a mature technology, optical communication has become ubiquitous and is growing as bandwidth demands rise [5]. However, with that ubiquity and growth, the total electrical power, and indeed the materials and complexity of systems, is also increasing at an extraordinary annual rate.

The design of optical communication systems is highly application dependant [1, 2, 6]. The high-speed, long-distance intercontinental links can use complex, power-hungry and expensive transmitters (Tx) and receivers (Rx). This is principally as the energy per bit (J/b) and cost per bit ($/b) can be kept low through long-term installations with high throughput and multiple end-users. As a direct comparison, hardware for applications such as VLC to mobile phones and personal computers, must be simple, low-power and cheap [3, 4]. This is especially true as smart-phones or personal computers imply a single end-user and a bandwidth low enough for simple or minimal forward error correction (FEC) schemes. It also implies battery operation with an expectation of long battery life and short product lifetimes (e.g. the 2-year average for mobiles). In terms of costs, VLC transmitters and receivers must: (i) be comparable to other system level components, i.e. memory, processor or display, and (ii) be suitable for high volume production, i.e. > 100,000 units.

1.1. Basic receiver operations and research directions

Taking a high-level approach, all optical communication receivers must perform eight basic operations [1, 2, 6]:

1. Convert incoming optical signals into an electrical signal, usually electrical current.
2. Amplify and convert the current signal into a form that can be processed easily.
3. Equalize and/or modify the signal to increase bandwidth or remove unwanted artefacts.
4. Demodulate the incoming signal, depending on its transmission modulation.
5. Sample the processed signal to recover the transmitted digital bit stream.
6. Output the signal in a form suitable for the end-user's application.
7. Optionally recover the clock embedded within the data stream, and
8. Optionally perform error correction tasks.

The role of the receiver front-end covers only tasks #1 to #6, however as part of the sampling operation many receivers also perform a clock and data recovery (CDR) operation [1], i.e. task #7. As error correction (task #8) requires a memory buffer, such tasks are separated from a receiver front-end and incorporated within encoding and protocol specific units within the overall system. The above front-end and back-end split can be viewed conveniently with respect to Ethernet technologies. The front-end Tx/Rx is often denoted as the 'PHY' as it relates to physical transmission and reception, it is also viewed as the physical layer in the Open Systems Interconnect (OSI) model. At the back-end, a Medium Access Controller (MAC) is used to implement the OSI data link layer, and contains both (i) transmit and receive buffers and (ii) framing, addressing and protocol level functions. Within the optical communication physical layer community, there are a number of key performance metrics that are driving innovation and the exploration of the current limiting factors [1, 2, 6]. These overall performance metrics are given below with current areas of innovation.

- Demand for increased bandwidth and transmission speeds
  - *Increase in bits per symbol through increased modulation complexity*
  - *Increase in channels per physical link though multiplexing in the time, frequency, phase or wavelength domains*
  - *Increased use of channel equalization techniques at both Tx and Rx*
- Reduction in energy usage and increased energy efficiency
  - *Decrease in the energy per bit*
  - *Trends towards quantum limited transmission (minimum optical power)*
  - *Increased transmission spectral efficiency for given real-world channel*
  - *Increased receiver sensitivity through (i) low-noise amplification, or (ii) optical and/or electrical coherent reception either homodyne or heterodyne detection*
- Decreased manufacture and design costs, and
  - *Increased Tx or Rx levels of CMOS integration, i.e. single-chip solutions*
  - *Decreased use of rare earth or high-cost materials*
  - *Minimization of bulky (for a particular CMOS node) analogue circuitry*
  - *Decreased design costs through Tx/Rx standardization*
- Decrease in bit errors and noise immunity
  - *Increased optical Tx extinction ratio*
  - *Decreasing Tx and Rx electrical noise*
  - *Decrease in inter-symbol interference*

1.2. Key receiver performance metrics

Before discussing the target application, a number of key receiver performance metrics need to be defined. These are (i) basic parameters at the communications level and (ii) measurements that can be made to assess the quality of a receiver, or its fitness for a particular application.

1.2.1. The data rate

The data-rate, R_D, measured in bits per second (b/s) is a measure of the serialized data throughput. As technology progresses, the throughput that is required at personal, commercial, national and international levels is increasing [5]. The data rate is intrinsically linked to the transmitter, channel and receiver bandwidth, BW, as each must pass all required frequency components in order to allow accurate, low-error decoding of the received signal back into binary data. In a communications receiver, the sampling point is chosen to be within the centre of the data symbol, thus allowing some protection from distortions on a signal's rising and falling edges. This leads to a lower frequency optimum bandwidth, BW_{-3dB}, which is related to the data rate [6], see Eq. (1). This is a mid-way point between (i) a low bandwidth that reduces input noise but produces low-pass inter-symbol interference, and (ii) a high bandwidth that captures all of a signal without distortion but also captures a large noise bandwidth and unnecessarily increases design specifications. For a 10 ns (100 Mb/s) data symbol, this would suggest a receiver bandwidth of 66.7 MHz. This is significantly below the knee frequency (Eq. (2)) of 250 MHz for a 2 ns rise time, T_R (10–90%) signal, i.e. 20% of a 10 ns 100 Mb/s data rate symbol.

$$BW_{-3dB} = \frac{2R_D}{3} \tag{1}$$

$$F_{KNEE} = \frac{0.5}{T_R} \tag{2}$$

1.2.2. The bit error rate

The bit error rate (BER) is defined as the number of single-bit errors within a continuous bit stream [1, 2, 6]. Data links at 100 Mb/s to 10 Gb/s may utilize error correction and are thus specified with 'native', no-FEC, error rates at 1×10^{-9}, i.e. a single bit error within a transmission of one billion bits. For links with slower data rates or where the application incurs significant channel distortions (such as VLC), the use of FEC may be a requirement. For these cases, the designer aims for the best native BER performance but as this cannot be guaranteed due to unknown factors impinging on the communications channel, a worst case maximum BER is chosen. For many FEC algorithms there is a limit as to effective recovery, at an approximate BER level of 1×10^{-3}. The BER is both a specification of the link and a measurable performance parameter. As an example, a BER of 1×10^{-6} to 1×10^{-9} may be specified as part of a protocol standard, but a measured value of 1×10^{-6} at a particular incident optical power may also become improved at a higher optical power.

There are two critical issues for the BER, (i) signal noise and (ii) inter-symbol interference. Assuming the use of non-return-to-zero (NRZ) on-off-key (OOK) modulation, the error rate is related to the probabilities that the noise of a 'one' or 'zero' cross a threshold, N_{th}. As both zeros and one are subject to noise, both have amplitude distributions about their means, (N_0 and N_1), these will have standard deviations of σ_0 and σ_1 respectively. The threshold and bit error rate can be approximated, for Gaussian noise profiles by Eqs. (3) and (4).

$$N_{th} = \frac{\sigma_0 N_1 + \sigma_1 N_0}{\sigma_0 \sigma_1} \tag{3}$$

$$BER = \frac{1}{2} erfc\left(\frac{N_1 - N_{th}}{\sigma_1 \sqrt{2}}\right) \tag{4}$$

Inter-symbol interference (ISI) is the bleed-through of symbols directly before and after, interfering with the reception of a bit. If we view the system as a low-pass filter of bandwidth, BW_{-3dB}, rectangular pulses become smeared in time with exponential rise and fall times [1]. As the signal is the supposition of multiple transitions the amplitude at the sampling point becomes corrupted by these tails. If the bandwidth decreases, these tails become longer in comparison to a bit period. Ideally, all ISI should have settled prior to sampling, however ISI acts to decrease the region within a symbol where robust sampling can occur.

1.2.3. Receiver sensitivity

For communications link budgets, it is useful to establish the receiver sensitivity [1]. This is defined as the incident optical power necessary for a receiver to reach the specified BER at the specified data rate, and is measured in dBm [1]. To ensure communication through a variety of signal attenuation factors, a high receiver sensitivity is required. We would therefore choose a receiver with a sensitivity of −20 dBm (0.01 mW) above a receiver with −5 dBm (0.32 mW). This would be particularly true if we knew the optical power was likely to be low at 0.05 mW due to perhaps low transmitter power and optical filtering. The sensitivity is dependent not only on the physical optical sensitivity, but also the received signal as a proportion of the receiver noise and inter-symbol interference. Thus, for a high-sensitivity receiver, (i) optical efficiency and electrical gain need to be high, while (ii) noise, inter-symbol interference and sampling phase-noise need to be low [1, 6].

1.2.4. The quantum limit

The receiver sensitivity can easily be reformulated in terms of the number of photons required per bit [1]. As this can be generalized to a theoretical receiver, we come across the theoretical ideal known as the quantum limit, QL. An ideal receiver would have zero noise (zero bits given by zero photon arrivals) and would receive single-photons with 100% efficiency. As photon arrivals follow Poisson statistics, the QL can be given by Eq. (5), where T_B is the symbol duration and $\gamma(T_B)$ denotes the average number of photons received per symbol. For a BER of 1×10^{-9}, approximately 20 photons are needed per symbol, which at 100 Mb/s ($T_B = 10$ ns), is 2×10^9 photons per second. For 650 nm (red) light this is 0.611 nW (−62.14 dBm).

$$\mathrm{BER}_{QL} = \frac{1}{2} e^{-\gamma(T_B)} \tag{5}$$

1.2.5. Energy per bit and power consumption

While the power dissipation of a receiver is a product of the supply voltage and receiver current, receivers with different data rates are often compared using the energy per bit, measured in Joules per bit (J/b) [1, 6]. As data rates are stretching beyond Gb/s, the energy per bit and therefore the total energy usage become significant issues [5]. Amplification and equalization circuitry often incur large energy costs; hence effort is being concentrated on efficient implementation. While reduced voltage supplies for integrated circuits (ICs) helps reduce power consumption, analogue design becomes more complex. Coupled with the bulkiness of robust analogue circuitry in advanced CMOS nodes (<45 nm), there is a trend towards denser digital circuitry in place of analogue where this is achievable, (i.e. trends of delay-locked loops (DLLs) replacing phase-locked loops (PLLs)). A significant open question is how to achieve high receiver sensitivity while obtaining a low energy per received bit? As transmitter efficiency and overall power/bit should also be included in the transmission energy budget of data, it is in our interest to reduce the power required of the most inefficient sub-system, along with improving its efficiency.

1.2.6. Integration and silicon area usage

The degree of receiver circuit integration, directly impacts system level costs and performance. Older receivers required separate ICs for the basic receiver tasks discussed in Section 1.1, however increased integration allows higher bandwidths, lower power dissipation and significantly reduced system level design times [7]. As an example, integrating a CDR circuit with the receiver removes high-bandwidth, accurate traces from a printed circuit board (PCB). It also reduces the power that must be dissipated by the input and output pads of multiple ICs, and drastically improves jitter [1]. Likewise, by integrating a photodiode in the same substrate as a CMOS receiver, the capacitance and inductance of bond pads and wires between an external photodiode and IC can be removed and thus achieving higher bandwidth. As multi-channel and multi-port ICs become more prevalent, some circuit blocks can be shared, thereby increasing area and energy efficiency. While requiring increased configurability, and difficulty in testing and verification, there is significant drive towards reduction in the number of input/output (I/O) pads within a system, hence integration of sub-systems onto a single IC.

1.3. Target application: visible-light communications

Visible light communication is a free-space method emerging as an alternative to local wireless systems [3, 4, 8, 9]. It is intended for data rates between 100 Mb/s and 1 Gb/s and has recently been ratified by the institute of electrical and electronic engineers (IEEE) (IEEE P802.15.7) [10]. A significant thrust is the dual use of energy-efficient light emitting diode (LED) technologies for illumination and communications [3, 10] (**Figure 1**). The advantages of VLC are principally

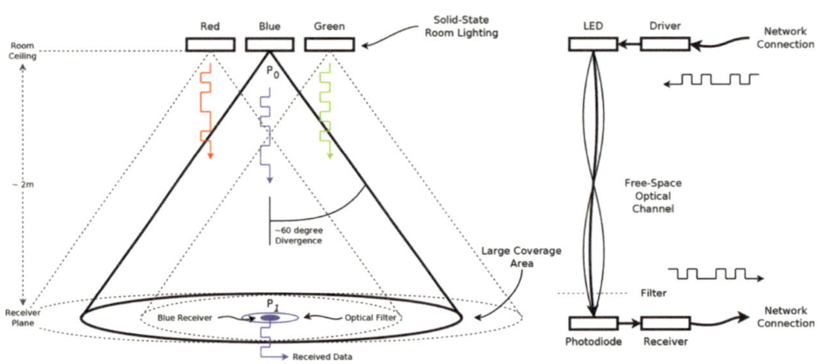

Figure 1. Visible optical communications and room lighting using red, green and blue modulated, LEDs. The wide divergence angle, presents an issue as the received optical power is low.

that the visible band is: (i) of wide bandwidth, (ii) is un-licensed and (iii) immune to electromagnetic interference from existing wireless technologies [1, 2, 6]. The primary target for end-users are mobile devices such as smart-phones and laptops where VLC can be used to supplement wireless systems for high bandwidth applications [3, 4] such as real-time, high-quality media streaming [5]. It is projected that as (i) the frequency allocation tables become increasingly dense and (ii) transmitters and receivers become increasingly difficult to produce at the high GHz radio levels, that some form of optical link—above the THz gap—will be required [3].

Despite advances in VLC, the application has a number of inherent issues that require application specific receiver designs. Firstly, the divergence angle of LEDs—particularly when wide-area lighting is needed—produces a large, diffuse illuminated area [3, 4]. When using a receiver IC of limited size, as per mobile systems, the receiver will capture a small proportion of the total transmitted optical power. Secondly, as optical sources are modulated at speeds close to their native bandwidth, both the average optical power and the modulation extinction ratio (the difference between high and low amplitudes) become smaller. Finally, in order to utilize the wide optical bandwidth, some form of optical filter is needed to perform wavelength multiplexing, i.e. multiple data channels on different colours. This has two implications, firstly that for ideally white illumination (wide-band), the receiver will receive a signal strength which is a narrow-band subset of wavelengths. A three-wavelength system (e.g. red, green, and blue, RGB), will naturally imply optical powers approximately one third of the total optical power. The second implication is that the filter itself will reduce optical power as they are neither perfect within their pass-band (signal attenuation) and have finite stop-band rejection (inter-colour interference and therefore finite separation). The effect of these issues, is that optical power incident on a receiver will be attenuated, while still being modulated at high data rates and with the user expecting both low error rates and low power consumption.

1.4. Research aims: extreme receiver sensitivity

We can consider VLC, as involving an inherent communications link budget issue, i.e. multiple attenuation factors necessitate reciprocal receiver gains. Further, the future requirements for low energy-per-bit communications would suggest an overall reduction in the link budget. The difficulty in such a scenario is twofold:

- First, the gain required in the communication budget may be extreme, especially if Gb/s links are required with multiple wavelength multiplexed channels, wide-area coverage and compact optically simple receivers.
- Second, as optical powers decrease and electrical gains increase, the signal-to-noise ratio (SNR) becomes a significant hurdle for robust detection. Johnson noise may limit analogue gain, especially if cooling is unachievable. Photon shot-noise may also become significant, without the possibility of using increased optical power at the transmitter (room lighting limited).

The work within this chapter therefore focuses on receiver designs with extreme sensitivity and high bandwidths to overcome the inherent attenuation factors of VLC, see [11, 12] and references therein. The overall method of using the avalanche multiplication of photo-generated carriers may use extra electrical power at the receiver—principally through increased biasing voltages. But, as this removes the need for analogue amplification circuitry, represents a more economical use of both electrical power and silicon area.

2. High-sensitivity optical to electric conversion

When electromagnetic radiation is incident upon a material, there are three processes that can lead to photon interaction, and thus the detection of that photon [13]. These are the (i) photoelectric effect [14], (ii) Compton scattering and (iii) pair production. Taking the high-energy processes first, we can begin to discount processes in order to arrive at the process whereby optical, and in particular visible, sensors operate.

- In Compton scattering, the incident photon is scattered by an atomic electron. It imparts some energy to the electron, meaning that the photon energy is reduced, and thus the wavelength becomes longer. Compton scattering therefore does not destroy the photon. As noted in [13], this effect is small for energies "below tens of KeV", i.e. 1 KeV is 1.24 nm and 10 KeV is 0.124 nm, and therefore can be discounted from visible applications within the band 400–700 nm.
- In pair-production, the photon energy is high enough to result in the production of an electron-positron pair, i.e. the electron's anti-matter counterpart. The photon energy must therefore be higher than $E_p = 2m_e c^2$, where m_e is the rest-mass of an electron. As this is 1.02 MeV, i.e. 0.0012 nm, this process occurs only for X-ray and Gamma-ray interactions, and thus can be discounted from any visible light application [13, 14].

Elimination therefore leaves the photoelectric effect—which is subdivided (within textbooks, patents, company websites, whitepapers and journals) into the external and internal photoelectric effects. It is therefore the only physical process suitable for optical to electrical conversion within visible applications [13, 14].

2.1. The photoelectric effect: converting light

In both photoelectric processes, a photon is absorbed by an atom, thus destroying the photon. If the photon energy is sufficient to overcome the work function of the material, a bound-free transition takes place whereby an electron is promoted from an outer electron orbital and is ejected from the surface [13, 15]. Remaining energy is accounted for by the kinetic energy of the electron as a free particle [15]. This is the external photoelectric effect as photo-electrons physically leave the material. The photon energy is given by Eq. (6), where h is the Plank constant, c is the speed of light, λ is the wavelength and v is the optical frequency.

$$E_P = \frac{hc}{\lambda} = hv \tag{6}$$

In contrast, for semiconductors with band-gap, E_G, between the valance, E_V, (i.e. outer-orbital bound electrons) and conduction electrons, E_C, (i.e. delocalized cloud of electrons) [16], a photon of energy greater than the band-gap ($E_P > E_G$) promotes an electron from the valence to the conduction band. As the absence of an electron in a valence state is described as a hole, the internal photoelectric effect produces an electron-hole pair [16]. This is a bound-bound or intrinsic transition [14, 15]. The electron is still ejected from the atom; however, not from the surface. If two electrodes are placed on the material with a slight potential gradient, or if that potential exists due to a p-n doped junction, the electron-hole pair are separated and drift due to their relative charges. With many photo-generated carriers within the material due to many incident photons, the bulk conductivity of the material increases [16], allowing a photocurrent to flow through an external circuit. Photons of high energy are highly likely to cause band to band transitions, however as the wavelength increases towards a photon energy close to the band-gap, the likelihood of transition decreases, given by the absorption coefficient, α. This leads to a long-wavelength cut off, λ_C, given by Eq. (7). For silicon, this is 1.1 μm, where the absorption coefficient is 1×10^1 cm^{-1}, whereas at 400 nm it is 1×10^5 cm^{-1}.

$$\lambda_C = \frac{1.24}{E_G} \tag{7}$$

The internal photo-electric effect is therefore the mechanism whereby semi-conductor materials can be utilized for the detection of light. As materials have various band-gap energies, different materials can be used to detect optical energy with different wavelengths [16]. Typically, less than one electron-hole pair is produced per absorbed photon, fundamentally limiting the internal quantum efficiency and the spectral responsivity measured in Amperes per Watt (A/W). The optically-induced current, I_P, flowing through a photodiode is given by Eq. (8) [16], assuming a detector thickness much larger than the light penetration depth, $(1/\alpha)$. q is the electronic charge, P_{OPT}, is the incident optical power, μ_n is the electron mobility. η is the quantum efficiency, ε is the electric field within the photoconductor, L is the distance between the contacts, and τ is the carrier lifetime.

$$I_P = q\left(\eta \frac{P_{OPT}}{hv}\right)\left(\frac{\mu_n \tau \varepsilon}{L}\right) \tag{8}$$

Within CMOS technologies, photodiodes are fabricated using p- and n- type dopants to form a p-n junction [14, 16]. As the average depth of absorption changes with photon wavelength, the depth of the junction is chosen, if possible, to maximize the received photocurrent. The width of the p-n junction is also critical in this; however, it also has implications for photodiode bandwidth. The response speed is restricted by three phenomena. Firstly, the capacitance of the p-n junction (dictated by the junction width). Secondly, the time delay of carriers generated outside of the junction, diffusing into the junction. And thirdly, the drift or transit time, t_r, of the carriers within the junction [1, 6, 16]. The transit time is given by Eq. (9), where L can be replaced by the junction width, W, as the electrode contacts are often fabricated on the top and bottom of the junction within planar technologies.

$$t_r = \frac{L}{\mu_n \varepsilon} \tag{9}$$

Eqs. (8) and (9) indicate both limits and opportunities with respect to receiver design. If the junction width or photoconductor length are reduced, the response speed increases, although this becomes limited if a thin junction width leads to a high capacitance and therefore a slow photodiode resistance/capacitance (RC) time constant. Reducing the width also reduces the photocurrent, requiring a high receiver gain to compensate. If the electric field is increased, both the photocurrent and diode bandwidth increase. It is here we encounter the fundamental limit and trade-off within high-speed, high-sensitivity receiver design. That transit time, and therefore bandwidth behaves in a reciprocal manner to photocurrent magnitude. As the electric field looks to be beneficial, this is a clue to guide research in the field.

2.2. The Avalanche gain mechanism: photo-carrier multiplication

A central challenge within receiver design is how to obtain the required amplification [16], or sensitivity. This must be done without undue use of (i) amplifier circuits (thus silicon area and power usage) or (ii) wide and therefore slow p-n junctions [1, 6]. Avalanche multiplication used in avalanche photodiodes (APDs), has been previously investigated for communications [1, 16–19]. This moves amplification into the diode providing initial gain without circuitry, and by increasing ε, allows both a larger photocurrent and shorter transit time.

The multiplication process uses an increased reverse bias voltage, V_R [16], creating an energy difference of $q(V_{bi} + V_R)$, where V_{bi} is the built-in potential between the p-side and n-side regions. The resultant increase in electric field, ε, accelerates a free carrier, labelled 1 in **Figure 2**, to a kinetic energy, E_K, sufficient to overcome the ionisation energy, E_G, of the material [16] (Eq. (10)). Upon a collision between a photo-carrier and the crystal lattice, the accelerated carrier ionises another carrier. An electron-hole pair, labelled 2 and 2′, is generated with those carriers then accelerated by the electric field, causing further ionisation [16, 17, 20]. This continues exponentially, creating an avalanche of carriers within the depletion region.

$$E_K = (1.5)\,E_G \tag{10}$$

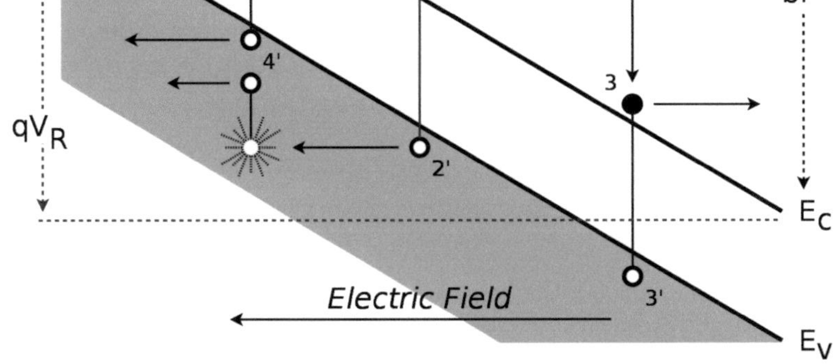

Figure 2. Band diagram showing reverse bias avalanche multiplication. When the electric field is elevated above the ionisation energy level, an accelerated carrier imparts a significant kinetic energy to a bound electron upon collision with the lattice. Adapted from Ref. [16], pp. 79 and 97.

APDs are biased such that multiplication achieves a constant generation rate, and thus constant gain [21] but not run-away avalanche leading to junction breakdown ([22] and references therein). The diode produces a current dependant on the incident photon flux with the gain and depletion region width being dependant on the bias [16, 21]. Some structures can be biased further into reverse bias, giving larger gains and greater sensitivities. In exceptional cases, run-away multiplication is used for specialised diodes called single-photon avalanche diodes (SPADs) [20, 22–25].

The use of avalanche gain, can be advantageous for sensitivity, as a gain of $M = 10–1000$, can be accomplished within the diode without the need of analogue amplification [1, 6]. Avalanche gain, is thus useful, however it comes with some disadvantages. Firstly, the higher bias voltage must be generated, necessitating additional power supply circuits or on-chip charge pumps. Secondly, multiplication is inherently random leading to increased noise. Finally, there is a gain-bandwidth trade off formed by the persistence of the avalanche process, whereby if the optical power decreases quickly, the avalanche caused by the previous high-optical power state takes some time to subside. To progress sensitivity and speed, through use of avalanche multiplication, circuits mitigating some of the above disadvantages are needed. For example, could we switch to smaller, lower power digital circuits and utilize very high multiplication factors by reducing the impact of multiplication noise?

2.3. Single-photon avalanche diodes: history and operation:

The Geiger region lies at the extreme end of the reverse photodiode range, beyond the linear avalanche gain region but prior to breakdown of a guard ring surrounding it ([11, 16, 21, 23, 25], and references therein). Initial research into avalanche behaviour, centered on microplasmas [21, 25]. These are small breakdown regions, corresponding to silicon defects [21, 25]. The historical study of microplasmas, resulted in artificial microplasmas [25] with guard rings to force a known breakdown region. These artificial structures later became known as Geiger-mode avalanche photodiodes (GM-APDs) or single-photon avalanche diodes (SPADs) [11, 23]. As they allow the avalanche mechanism to run-away, breaking down the p-n junction, their gain factors can be greater than 1×10^6, i.e. a single-photon, yielding a single electron-hole pair is able to produce a sizable avalanche photocurrent.

The Geiger region provides long periods of quiescent operation at a voltage of $V_{OP} = V_{BD} + V_{EX}$, where V_{EX} is the excess bias applied above the breakdown voltage, V_{BD} [20–23]. Upon the arrival of a photon, the device transitions from the quiescent point (**Figure 3**), to the I-V curve, with the avalanche quickly building up (10–100 ps). The current flows through any external series resistances, causing a significant voltage drop, designed to be of magnitude V_{EX} over that resistance. This reduction in voltage, (i) is sensed, indicating a single-photon event and (ii) reduces the voltage across the SPAD p-n junction [22, 23]. The reduced voltage, decreases the kinetic energy of carriers and halts the avalanche, a process known as quenching. The lower current flow allows the SPAD to recharge to $V_{BD} + V_{EX}$. There are two principal operating modes, passive quenching with passive reset (PQPR) and active quenching with active reset (AQAR) [23]. The passive circuit uses a large ~100 kΩ resistance (typically implemented using transistors), R_Q, and optionally a smaller 50 Ω sense resistor. The circuit recharges with a time constant given by $R_Q C_{SPAD}$, where C_{SPAD} is formed by (i) the p-n junction

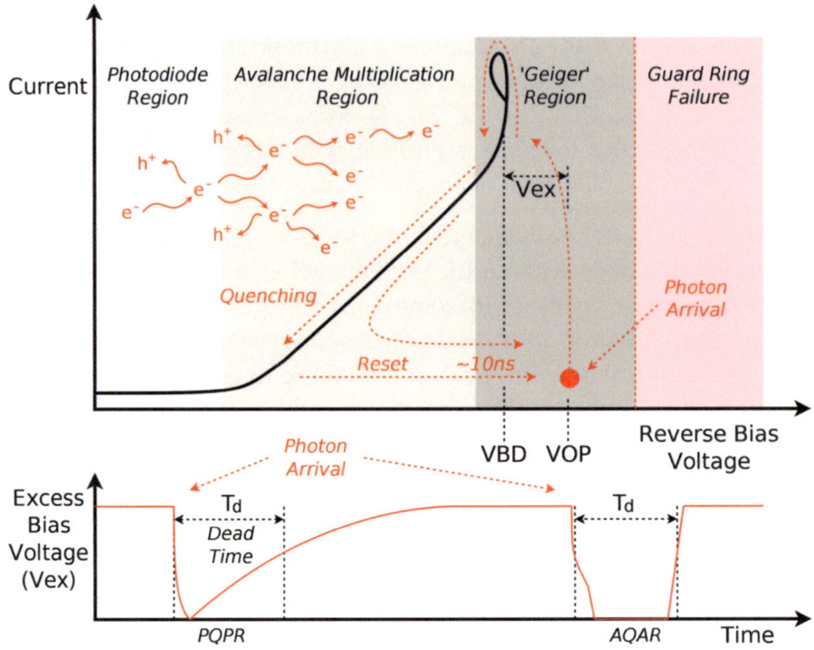

Figure 3. Top: the bias regions and I-V curve of a SPAD, showing the avalanche, quench and reset cycle. Bottom: the voltage-time pulses for both passive and active quenching circuits. Notice the steep, well defined edges of the AQAR pulse shape. The dead-time T_d is shown for both.

capacitance, and any parasitic capacitances on the diode (**Figure 4**). In the active case, the change in voltage triggers a quenching transistor which pulls V_{EX} to zero. After a short delay a separate low on-resistance transistor resets the device to V_{EX}.

In **Figure 4**, both positive-going and negative-going PQPR SPAD circuits are shown, along with the parasitic anode to substrate (C_{AS}) and cathode to substrate (C_{CS}) capacitances. The transistor gate voltage, DT_{ctrl}, can control the on-resistance, allowing adaptation of the SPAD dead-time, T_d. The n-type metal-oxide-semiconductor (NMOS) transistor circuit is preferable

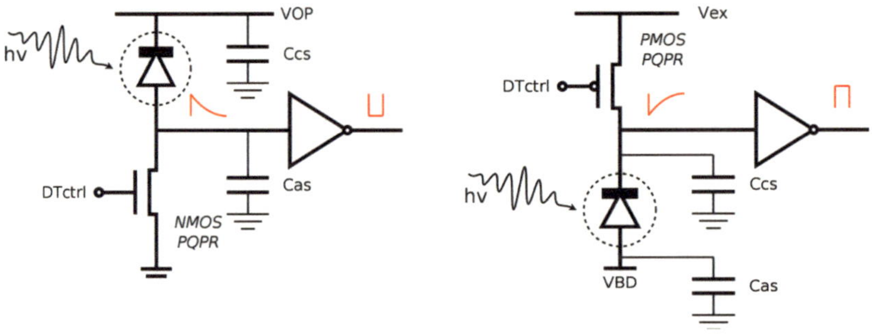

Figure 4. Two PQPR circuit topologies. Left: positive-going circuit with an NMOS quenching transistor and a small anode parasitic capacitance on the diode moving node. Right: negative-going circuit with a PMOS transistor, but with a larger cathode to substrate parasitic capacitance.

as (i) C_{AS} is typically smaller than C_{CS} and (ii) NMOS transistors can be integrated within a SPAD array without the deep N-well required for p-type metal-oxide-semiconductor (PMOS) devices.

It is with circuits such as this, that high-sensitivity receivers can be realized [11, 12]. One circuit approach would be to combine avalanche gain exceeding that used in APDs, with topologies known to give high sensitivity. The SPAD circuit can be used to replace the continuous-time photodiode, trans-impedance amplifier (TIA) and feedback network, Rf, circuitry (**Figure 5**, left). A small digital inverter, necessary to prevent loading on the SPAD, replaces any post-amplification, while the resulting full logic swing can be used with a digital counter (**Figure 5**, right). This discrete time "direct-to-digital" approach allows all gain to be implemented within the diode, removing complex analogue circuitry, and drastically reducing area requirements. Electrical power is expended within the SPAD due to a total voltage of V_{OP}, and the avalanche current flow per detection, however analogue amplifier circuits typically require continuous bias currents for each amplification stage branch [1, 6].

2.4. Single-photon avalanche diodes: key performance parameters

Before discussing SPADs within optical receivers, there are several key performance parameters that must be discussed. As the field is still under active investigation, there are as yet no definitive methods to bring low-level parameters up to the communications level, particularly the bit error rate (BER). This contrasts with mature photodiode receivers where the receiver BER can be derived [1, 2, 6]. Taking the properties of a single SPAD first, we encounter non-ideal behaviour away from true single-photon counting [23]. For receivers, the concept of the receiver power penalty within the link budget [6], can be used to estimate the effects, to first order, of various offsets and non-idealities. The power penalty, expressed in decibels (dB), models the adjustment in optical power required to compensate for an effect.

1. SPADs produce avalanche events despite the absence of light, this is the so-called dark count rate (DCR), measured in counts per second (cps) or Hertz (Hz) [20, 21, 23, 24]. This is related to (i) thermal generation of carriers, and (ii) tunnelling of carriers predominantly due to traps within the band structure. The DCR has steadily improved through (i) academic investigation of guard ring structures such as retrograde guard rings [26], (ii) modelling and understanding of the transient behaviour of the junction itself [27] and (iii) CMOS process cleanliness, and can now be as low as 1 Hz [27]. A per-SPAD rate such as this is now effectively ideal in

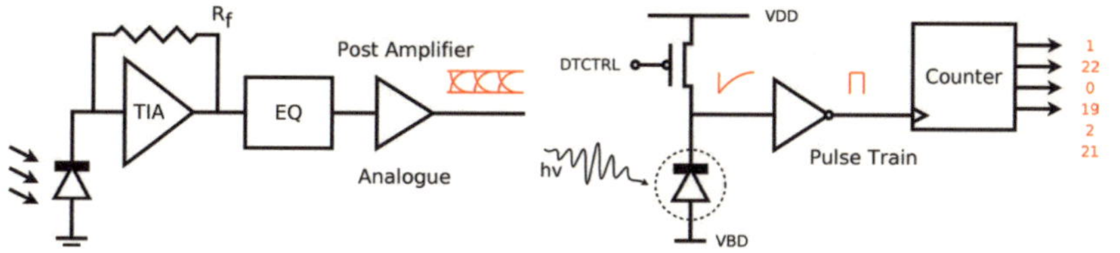

Figure 5. Left: a continuous-time conventional photodiode receiver, and right: a discrete-time replacement with single-photon sensitivity, where the SPAD and inverter allows optical to electrical conversion into a voltage (i.e. trans-impedance) with full logic swing magnitudes.

comparison to high (>100 Mb/s) data-rate communications. The DCR limits the optical photon flux at which reliable single-photon detection can be achieved. Therefore, the incident optical power would need to be increased by a power penalty, PP_{DCR}. This is given by Eq. (11), where DCR_{PC} is the DCR per channel per symbol period as a proportion of the required number of photons per symbol, similar to the voltage threshold offset power penalty ([1], p. 71). As the DCR and symbol period decrease, this penalty becomes negligible. For a DCR_{PC} of 10% of the required photons (~20), the PP_{DCR} is 0.79 dB.

$$PP_{DCR} = 10\log_{10}\left(1 + \frac{2\,DCR_{PC}}{100}\right) \tag{11}$$

2. A SPAD—like a photodiode or avalanche diode—does not detect light with 100% efficiency. The photon detection efficiency (PDE), typically some 20–40%, is linked to three processes, (i) the internal quantum efficiency [16], (ii) the wavelength specific absorption depth and the width and depth of the p-n junction, and (iii) the avalanche turn-on probability [20, 23, 25]. This latter factor is a product of the kinetic energy achievable before a carrier recombines, and is thus dependent on the electric field. In combination with the DCR, this limits the lower optical power that can be reliably detected. A power penalty, PP_{PDE}, to account for the finite detection efficiency can be calculated using Eq. (12). For a SPAD with 20% PDE, this is 7 dB.

$$PP_{PDE} = -10\log_{10}\left(\frac{PDE}{100}\right) \tag{12}$$

The PDE has increased through commercial R&D into SPAD-dedicated micro-lenses. However, it has proved difficult to improve the internal quantum efficiency as increased electric fields correspondingly increase the DCR, while wider depletion regions require modifications to dopants that may not be an option for a multi-customer CMOS foundry. Correspondingly, the drive for high-PDE has led to the optimisation of the layer stack above the silicon surface, including reduced metallisation, thinner nitride layers and anti-reflection coatings, or the use of back-side illuminated structures. The combination of these design options has promoted foundries such as ST Microelectronics and TowerJazz to develop dedicated optical CMOS processes for SPAD implementation.

3. The dead time, T_d, is a discharge and recharge period, during which the bias falls below the normal $V_{EX} + V_{BD}$ [20, 21, 23]. No photon arrivals can be detected, although for PQPR as the voltage exponentially recharges there is an increasing probability of an avalanche event, and thus the dead time can be extended, an effect called paralysis [28, 29]. A short dead time is advantageous as it maximizes the detections per second, C_{MAX} (Eq. (13)). Unfortunately, an arbitrarily short dead-time, cannot be achieved as second order effects become more pronounced [23]. The dead time has been reduced within the literature, principally through the use of AQAR, and the use of smaller diodes. For AQAR, the maximum count rate and the dead time required for a target minimum number of photons per second, $N_{PHOTONS}$, for a number of SPADs, N_{SPADS} and target number of channels, N_{CHANS}, are given by Eqs. (13) and (14) respectively. For passive quenching, both equations are multiplied by 1/*e* [28, 29].

$$C_{MAX} = \frac{N_{SPADs}}{N_{CHANs}\, T_d} \tag{13}$$

$$T_d = \frac{N_{SPADs}}{N_{CHANs} \lfloor N_{PHOTONs} \rfloor} \tag{14}$$

4. Without a long dead-time, there is a small but finite probability of a secondary avalanche, denoted as an after-pulse, ap [20, 23]. This is related to the release of a charge carrier from a trap within the device. The number of trapped carriers is dependent on the charge flow during an avalanche, hence reducing SPAD capacitance is beneficial [22, 24]. The number of traps is a function of process cleanliness and device structure, while the trap lifetime, and hence the approximate required dead-time is a function of temperature. A minority-carrier effect, may limit the lower-bound of the dead-time with respect to after-pulsing, and hence fundamentally limits reductions in dead-times [22].

5. The Poisson distribution of photon arrivals is complemented by a similar timing variation in photon detection [20, 26]. This event timing jitter is a product of two contributions. Firstly, the statistics of avalanche build up [30], lateral spreading of the avalanche throughout the junction volume and the carrier transit time [31]. Secondly a lengthy, but low-probability jitter tail is caused by carriers deeper within the substrate diffusing towards the p-n junction. The use of silicon-on-insulator (SOI) CMOS processes may offer some mitigation, however a significant trend for jitter reduction has been SPADs of small diameter and the tailoring of the SPAD doping.

Arrays of SPADs in CMOS planar processes also have a number of non-ideal behaviours and limitations. As the SPAD dead-time or IC application may necessitate the use of multiple diodes, array level limitations have a large impact on final receiver performance [11].

6. The fill factor (FF) is the ratio of the optically receptive area to the total device area and is a key parameter in obtaining the goal of a high sensitivity receiver. For a square array of circular SPADs, the fill factor (FF_{SQ}) is given by Eq. (15). Where A_{ACT} is the total optically active area, A_{ARR} is the total square array area and N_{SPAD} is the number of SPADs within the array. D_R is the radius of the per-SPAD active area, D_G is the guard ring width, and D_S is the guard-ring to guard-ring separation. The fill factor significantly alters the receiver sensitivity as photons incident on inactive regions are not counted but non-the-less are included in link budgets. This leads to a fill-factor power penalty, PP_{FF}, given by Eq. (16) [11]. The fill factor is often limited by the diode geometries available for a given noise or capacitance specification [24], the complexity integrated onto the receiver and the CMOS process design rules [22, 24].

$$FF_{SQ} = \frac{A_{ACT}}{A_{ARR}} = \frac{\pi D_R^2 N_{SPAD}}{\left[2 D_R \sqrt{N_{SPAD}} + (2 D_G + D_S)\left(\sqrt{N_{SPAD}} - 1\right)\right]^2} \tag{15}$$

$$PP_{FF} = -10 \log_{10}\left(\frac{FF}{100}\right) \tag{16}$$

Prior to 2010, the fill-factor was often limited to 1–2% [11]. This necessitates a factor of 50–100 increase in optical power (PP_{FF} = 17 dB), to return the detected photons per data symbol to the ideal ($FF = 100\%$) theoretical value. Despite restrictions [24], the fill factor has been increasing through a number of techniques. N-well sharing, out-of-array electronics, reduced quenching circuit areas, NMOS-only logic, increased diode radii with retained low-noise, use of clustered mini arrays, non-circular geometries and micro-lenses, have all lead to fill factors reaching recent maximums of 67% [32] and 70% [33]. With a fill factor such as 70%, the optical power would need to be increased by a factor of 1.43 ($PP_{FF} = 1.55$ dB). To date, the fill-factor has been a particular focus of both academic and commercial R&D with several companies becoming heavily invested.

7. Detector cross talk, which is often at the 1–2% level in SPAD arrays using out of array circuitry, is induced by secondary photons in the substrate [34]. These are produced through radiative recombination during an avalanche. In-array circuitry has also been recognised as a contributor [22]. Therefore, while area within the array should be used to maximise the functionality to area cost ratio, high-speed circuitry, clock buffers and digital supplies should be placed outside the active areas. Cross talk can be reduced with increased SPAD spacing; however, this negatively impacts the fill factor.

8. As with all sensors, the temporal response departs from the ideal as characterized by the step and impulse response [11]. Upon the reception of a "zero" to "one" bit-level transition (a positive step assuming OOK), the ideal would be for the received transient to be limited by the channel. The step response however may cause a finite rise time and ripple that must settle before the output can be sampled. Likewise, upon the reception of a "zero" after a "one", the impulse response limits the speed at which the receiver settles to the new level. Upon either transition, the system must wait before sampling, and thus the symbol rate is fundamentally limited (discussed further later).

3. Circuit approaches and topologies

To date there are three distinct circuit topologies used within SPAD-based receiver designs. Two of the topologies use the innate full-logic-swing of the SPAD to produce an all-digital approach with either a sampled integrating or continuous time output. The third topology again aims for a continuous time output, but does so using the SPAD to switch current steering circuits. Each topology has advantages and disadvantages; however, the largest impactor is the level of integration within a CMOS technology as a function of the receiver functions discussed in Section 1.1. Along with these topologies, pure analogue, parallel connections have been formed into so called multi-pixel photon counters (MPPCs), these differ from SPAD arrays as the diode is not used to create a full logic level voltage. A positive feasibility study of MPPC use for communications can be found in [35].

3.1. Digital count summation: digital synchronous discrete time

One of the first all-digital multi-SPAD arrays custom designed for optical communications [12], aimed to replicate a synchronous integration mode-receiver [1]. A receiver of this type is presented in Section 3.2 as an example SPAD-based receiver. Integration prior to signal

http://dx.doi.org/10.5772/intechopen.68935

sampling allows: (i) the output signal amplitude to be increased, i.e. allowing a doubled integration time, will approximately double the output amplitude, and (ii) noise which has a period shorter than the integration window is reduced. As SPADs produce a discrete time pulse, integrating signals is equivalent to pulse counting [11].

The topology is shown in **Figure 6A**, where the pulses from a SPAD are counted digitally using a ripple or synchronous counter. An integration window (**Figure 6B**) shorter than the symbol period can be used (i) to allow rejection of photon counts caused by the rise and fall periods of the input, (ii) to reject DCR events outside of the window, and (iii) to stop, sample and reset the digital counter. The addition of multiple high-speed synchronous data streams from multiple 'pixels', (formed from a SPAD and a counter), can be performed easily and with low latency. A multi-diode approach therefore allows multiple detection events during periods equivalent to the SPAD dead-time. A SPAD array output, in the form of a wide bit-depth synchronous digital stream can then either be (a) output to allow array diagnostics and signalling formats such as pulse amplitude modulation (M-PAM) or (b) thresholded on-chip allowing recovery of single-bit serial data streams. As the readout clock can be adjusted in phase relative to the transmitter clock, the position of the integration window can be chosen to minimize the BER, and its width can be optimized to maximize received signal amplitude.

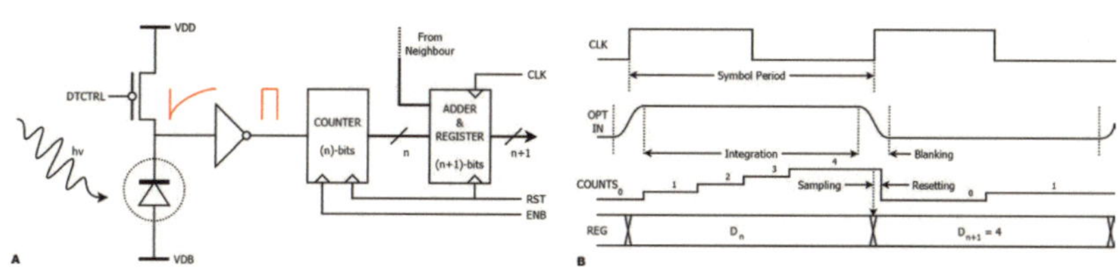

Figure 6. (A) The circuit topology with a binary counter per SPAD with synchronous binary addition of multiple SPAD counts within a defined integration window. (B) Timing diagram showing an input optical symbol, with an integration period blanking off the low-high and high-low transitions.

3.2. XOR-tree summation: digital asynchronous discrete time

To decrease the required logic, negated AND and negated OR (NAND/NOR) trees were investigated for combining multiple SPADs in small, higher fill factor mini- silicon photomultipliers (SiPMs) ([11] and references therein). However, to operate at high counting rates, the SPAD pulse (10–30 ns) must be reduced to <1 ns using per-SPAD pulse shortening. This is still smaller than per-SPAD digital counters, however it prompted the investigation of exclusive-OR (XOR) trees for mass digital pulse summation of larger SPAD arrays [36, 37] (**Figure 7A**). The XOR truth table lends itself readily to a summation task as sequential leading edges of SPAD pulses (without pulse shortening), simply toggle the output (**Figure 7B**). The limitation of this is the rate at which the final XOR within a tree can toggle its output.

As XOR gate standard cells occupy a smaller area than the digital logic required per SPAD in both digital summation and pulse-shortened NAND/NOR summation, the fill factor of the circuit can be increased significantly, in the case of [36], a fill factor of 43% was achieved. Indeed, such a receiver has demonstrated 4-PAM VLC, with highly discernible Poisson-limited levels,

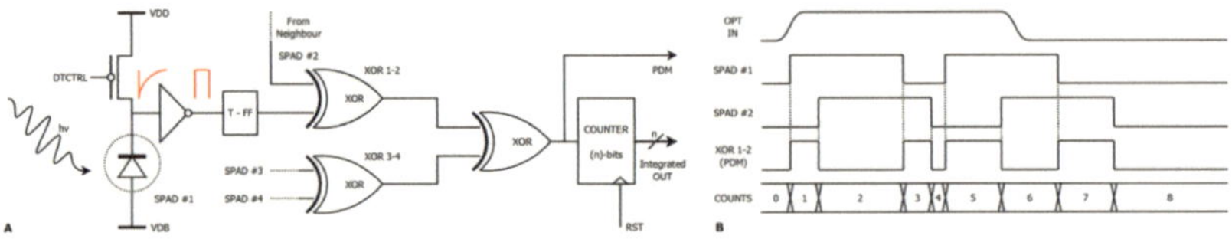

Figure 7. (A) The XOR summation of pulses, without pulse shortening, of four SPADs. The toggling of the output is a form of pulse density modulation; however, a binary counter can be used for pulse integration during a symbol. (B) Timing diagram of the XOR circuit for events on SPADs 1 and 2.

albeit with a low data rate [38]. As noted by the authors, this topology can be combined with (a) the digital summation approach, (b) on-chip modulation decoding logic, (c) signal processing logic, and (d) the adaption of the XOR tree and number of SPADs per channel. This would allow receiver optimization for power usage and optical sensitivity [38].

3.3. Current steering summation: analogue continuous time

In **Figure 8**, the current summation topology is shown [39], this uses the SPAD to switch currents onto output nodes shared by multiple SPADs, IQ– and IQ+. When the SPAD toggles, the current path changes due to the opening or closing of transistors MS1 and MS2. Transistor MCS could be implemented outside of the array, and transistors MS1 and MS2 can be far smaller than large digital circuits such as synchronous carry look-ahead counters. The authors therefore hoped for a high fill factor, however the use of PMOS devices limited fill factor through the SPAD N-well spacing rules. A fill factor of 3% was achieved however this was not optimized. A deeper issue is that the topology includes an inherent signal attenuation, whereby a full-logic swing on the SPAD (V_{EX} = 1.6 V) and inverters results in a maximum per-SPAD steered current of 100 uA. With a value of 300 Ω for RO+ and RO– this yields 30 mV per SPAD. Larger trans-impedance values could be used (e.g. 1 kΩ, yielding 0.1 V per SPAD), however trans-impedance bandwidth decreases as this resistance increases, i.e. $BW_{-3dB} = (2\pi R_O C_D)^{-1}$ [1]. It seems a shame to insert (i) a signal attenuation, and (ii) the transform {SPAD avalanche current to SPAD voltage to readout current to readout voltage}, into the topology to read out the SPAD array and communication signals.

While summation using Kirchhoff's current law for many SPADs is more efficient in terms of area and power than a multi-bit digital summation tree, and thus a lower energy per bit,

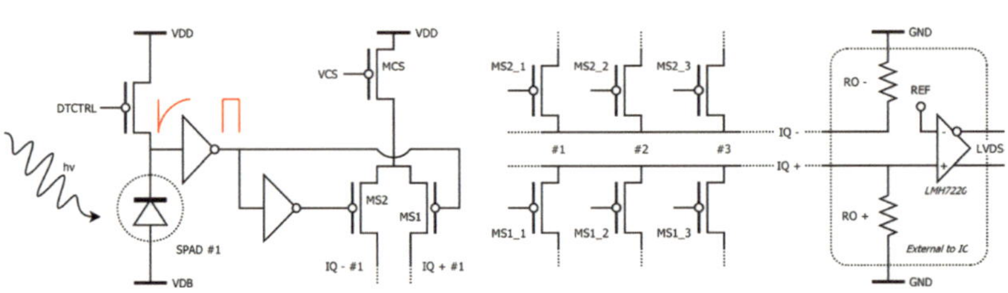

Figure 8. A current summation topology where a SPAD steers current through transistors MS1 or MS2. Multiple outputs can be added using Kirchhoff's current law on nodes IQ– and IQ+. Off-chip trans-impedance resistances RO– and RO+ are used to convert the current to a continuous-time voltage.

http://dx.doi.org/10.5772/intechopen.68935

this readout structure introduces noise through transistor mismatch and the Johnson noise of the trans-impedance resistances. As this noise acts upon the voltage levels of "zeros" and "ones" at the analogue LMH7220 comparator, the steered current per SPAD, and thus the per-SPAD voltage has a fundamental limit, especially if the receiver is intended for M-PAM modulation schemes. This is an additive readout noise in combination with both Poisson photon arrival and noise induced by dead-time, DCR and after-pulsing effects. Further, to compare the energy per bit and the receiver area, modelling would need to include the power dissipated and area of (i) both RO resistors, (ii) the comparator, (iii) the field programable gate array (FPGA) serial-deserial (SERDES) input pads and (iv) the latching flip-flop. This would be the case as this yields the same synchronous full-logic level digital bit as the previous topologies, albeit single- rather than multi-bit. While the output is asynchronous with samples taken when the data eye is maximally open, with no blanking periods SPADs breakdown due to photons within the transitions of the optical source, thus preventing them from contributing once the optical signal has reached its correct level. This would be critical for reducing the standard deviations, σ_1, σ_2 and σ_3, upon M-PAM levels with means N_1, N_2 and N_3. Modulated OOK data streams have been received with this topology with 125 Mb/s at a BER of $1 \times 10^{-3.5}$ achievable with a dead time of 8.8 ns and before decision feedback equalization [39, 40]. We must note however, that to be fair all receivers should be compared without equalisation, although equalisation represents a logical subsequent signal processing step for low BER communications.

3.4. An exemplar receiver IC

To demonstrate the concepts and to investigate issues inherent in the design of such receivers, the digital summation topology has been investigated using a 32 × 32 array of 8 μm SPADs in 130 nm imaging CMOS (**Figure 9**) [11, 12]. This does use a hybrid approach with NAND/NOR signal addition prior to the main digital parallel summation tree, but does so mildly using 'super-Pixels' formed from four SPADs (**Figure 10**).

The design, discussed in detail within [11, 12], operates as an integration-mode receiver, with two blanking periods of 500 ps at the start and end of a symbol. These screen out SPAD events during the transitions of an optical source, and thus decreasing the variability of counts within an integration period. The digital summation tree allows clocking at 100 MHz, culminating in either a single channel 14-bit output or a 9-bit 16 channel output operating at 25 MHz per channel. As the receiver was intended as a test-harness for various concepts, the array treated the fill-factor and PDE as fixed offsets that could be improved in subsequent generations. Rather than designing to maximize known performance metrics, the array was designed to test transient issues along with unknown factors important for both communications and high-sensitivity, high sample-rate metrology.

By passing full logic swing synchronous parallel data, rather than (a) fast toggling pulse density modulated data, (b) asynchronous ripple-counter data or (c) currents requiring a trans-impedance, the receiver requires no circuitry prior to its test data being captured by an FPGA FIFO, or being thresholded. The receiver, summarized in **Table 1** (see [11] for results/modelling), represents a test harness with dedicated modes for array characterization.

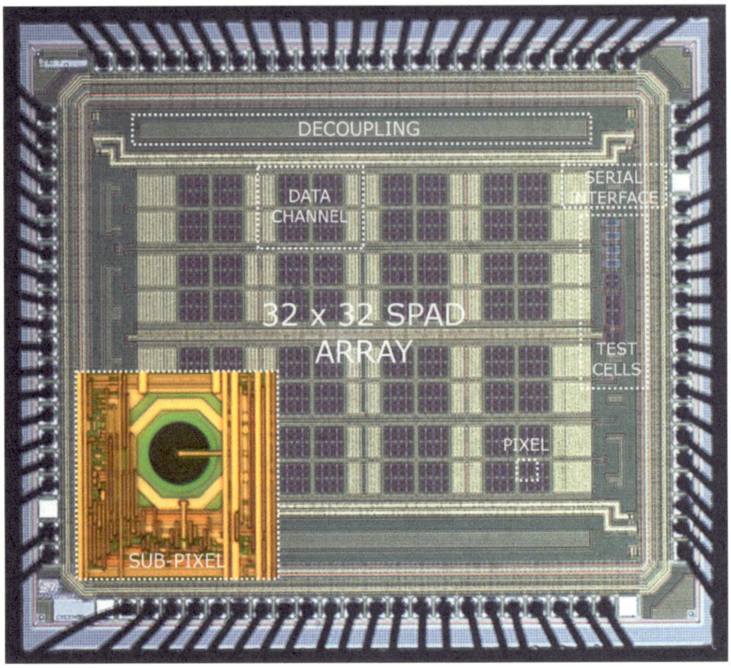

Figure 9. Micrograph of the 1024 SPAD digital summation integration-mode receiver with the SPAD inset. This has a fill factor of 2.4%, and implements up to 16 'Data Channels' each of 64 SPADs.

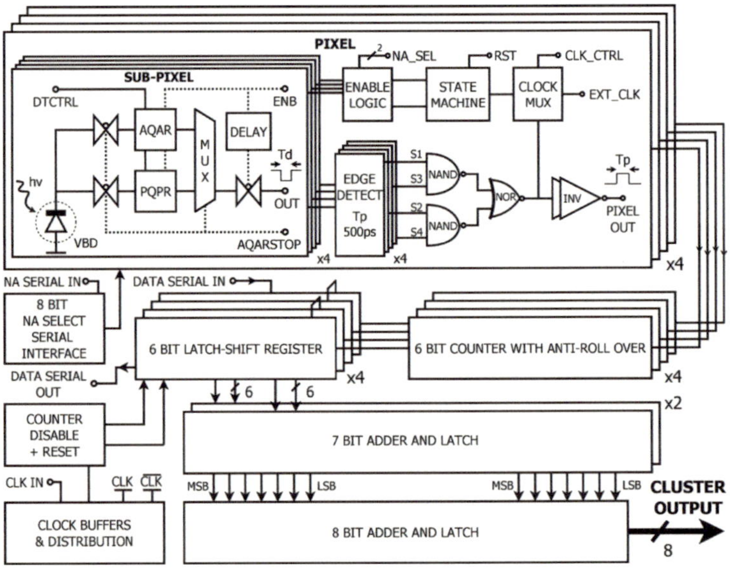

Figure 10. Block diagram of the architecture for 16 SPADs, showing per-SPAD active and passive quenching, dead time control via the analogue voltage DTCTRL, and pulse shortening. A binary counter (6 bit) is used per group of four SPADs and NAND/NOR signal addition is used.

Parameter	Value	Units	Notes
CMOS process	130 nm Imaging	–	90 nm metallization
Peak PDE	20	%	VEX = 1.2 V at 450 nm
SPAD VBD	−13.05	V	–
Dead Time	5–250	ns	Active or passive
Pulse shortening	500	ps	For NAND/NOR Sum
SPAD VEX	1.35	V	Typically, 1.25 V
Fill factor	2.42	%	Exc. I/O Pads
# SPADs	1024	–	Circular
SPAD diameter	8	μm	Active area
Quenching	AQAR or PQPR	–	Adjustable
Digital VDD	1.2 and 3.3	V	Core/IO
Pixel counter bits	6	Bit	Per 4 SPADs
Output width	14 or 9	Bit	Single- or Multi-Channel
# Channels	1 or 16	–	Configurable
Die size	2.4 × 2.1	mm	~5 mm^2
Clock frequency	108 100	MHz	Pad-Limited (Max) (Typical)
Median DCR	2.5	kHz	16°C, V_{EX} = 1.35
Max count rate	65	Gphoton/s	~10 ns Dead time
After-pulsing	0.9	%	1 μs window
Energy per bit	370	pJ/Hz	Estimated
Sensitivity estimate	−31.7	dBm	100 Mb/s BER = 1×10^{-9} 100% extinction ratio
Sensitivity estimate calibrated	−28.5	dBm	Inc. LED extinction ratio reduction

Table 1. Performance details for the 32 × 32 digital summation integration-mode receiver.

4. Results

This section will present experimental results from the example digital summation integration-mode receiver. The sensitivity curve or photon transfer curve will be shown and a model for this will be presented. The step response will also be shown as it makes a direct impact on communications performance. This will also be modelled illuminating the choices a designer has with respect to the SPAD dead-time, number of SPADs, digital-counter read-out rate and maximum modulated input power. Finally, experimental communications results will be shown, demonstrating both feasibility and several issues we have discussed.

4.1. Photon transfer curve and optical sensitivity

In **Figure 11**, the photon transfer curve (PTC) is shown. This plots the detected photon events per second against the input optical power. The N_a modes describe a time multiplexing approach, which while beneficial for the temporal response, will not be discussed here (see [11]). The N_a = 4 mode is of interest as it turns on all SPADs within the summation. At the centre of the curve, the receiver has a linear response, however it saturates at both the low and high levels due to the combined DCR and the active or passive quenching maximum count rate, C_{MAX}, respectively. Passive quenching shows both the presence of the 1/*e* factor in its maximum counting rate, and the issue of paralysis whereby the count rate starts to decrease at high optical powers. For communications, the receiver must operate within the linear region, as nonlinearities act to modify the extinction ratio of the detected signal, and thus negatively impacts the BER.

For design of SPAD-based receivers, active quenching appears to be advantageous, however as it requires extra circuitry per-SPAD, the choice of active or passive quenching must be taken in view of top level communications performance metrics. By reducing the dead-time, a higher count rate is achievable, while an increased number of SPADs has the same effect but comes with increased total DCR. It is advantageous to reduce the DCR through both innovation and by turning high-DCR SPADs off (as shown on **Figure 11**) as this extends the lower optical power limit and pushes the receiver closer to being quantum limited.

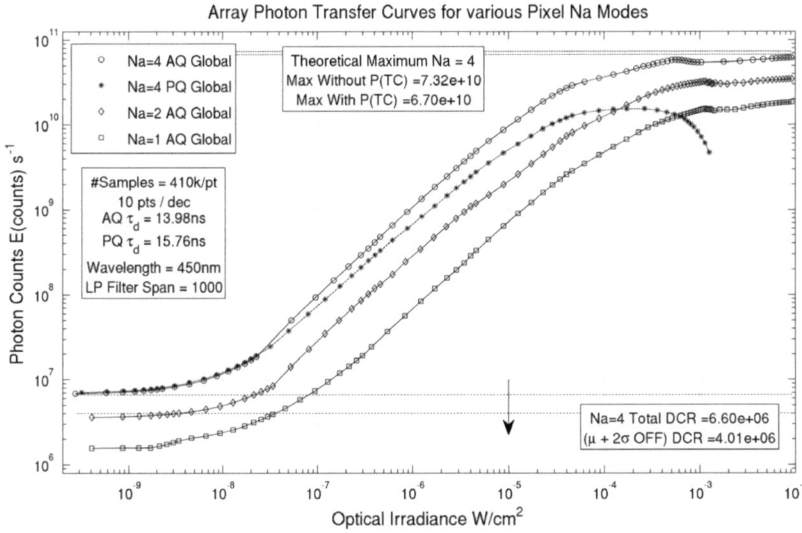

Figure 11. The photon transfer curve of photon counts per second against optical input power. The curve is linear within the centre; however, saturation occurs due to the combined DCR of all SPADs within the summation and the active or passive quenching maximum count rate, C_{MAX}.

The dynamic range can be calculated as: (i) the ratio of maximum (i.e. C_{MAX}) to minimum (i.e. $\sum(DCR)$ observable signals (approx. 80 dB in Figure), or (ii) the ratio of the maximum to the noise of the dark count floor rather than its absolute value, i . e. $\sqrt{\sum(DCR)}$ (approx. 149 dB). If we accept that between two modulated amplitude levels, a suitable gap must be left in order to achieve the required BER, the size of the linear region gives an estimate as to the theoretical number of M-PAM levels that could be implemented. With the required offsets, due to the finite fill factor and photon detection efficiency known through measurement of the PTC, other power penalties can be assessed and ranked as a guide to future design.

4.1.1. A competitive count model of SPAD avalanche initiations

The PTC can be modelled using competitive interaction, matching the measured experimental data (**Figure 12**). The arrival of a photon, (assuming AQAR), predicates loss of a photon, dark count or after-pulse if one occurs during the dead time. This creates competition at high count rates, in which (i) photons prevent dark counts or after-pulses, (ii) dark counts prevent photon counts or after-pulses, and (iii) after-pulses prevent photon counts or dark counts.

A simple model, describing the number of output counts as a function of input photons is given in Eq. (17). Where FF is the fill factor, DCR is the dark count rate, ΔT represents the time that counts are collected and $P(ap)$ is the after-pulsing probability. $PDE(\lambda)$ is the photon detection efficiency at a particular wavelength and $\psi(\lambda)$ is the number of photons, of wavelength λ, incident per second. This does not account for the dead-time, although it could be combined with hybrid paralysable models of dead-time limited counting [28, 29]. This would modify the output counts, N_{counts}, but would not include the concept of competition due to events other than photon arrivals. The model assumes that photons, dark counts and after-pulses can be separated, which while achievable with careful experimentation, cannot be performed during normal operation. The question therefore arises of what is the DCR level when the receiver is being illuminated, knowing that some DCR events will not be detected as the SPAD has already broken down? As noted within [23], a photon arrival and an after-pulse, or a dark count and an after-pulse are not statistically independent.

$$N_{counts} = [1 + P(ap)](PDE(\lambda)\psi(\lambda)FF + DCR)\Delta T \tag{17}$$

As high counting rates prevent the separability, independence and static contribution assumptions being used, the concept of an enable time, T_{enb}, as a proportion of ΔT, can be introduced, i.e. an avalanche induced dead time will remove a period, T_d, from ΔT. The time in which a SPAD is in its receptive state, is then given by Eq. (18), and is the first step towards a count competitive model. The values ψ_{act} and DCR_{act} are the actual numbers of photon and dark detections, however these are intermediate output values in Eq. (18) as they must be corrected for count competition [11].

$$T_{enb} = \Delta T - [(1 + P(ap))(\psi_{act} + DCR_{act})(T_d)] \tag{18}$$

A regression can be made backwards to the input values for photons, ψ_0, dark counts, DCR_0, and after-pulses, AP_0, that can be calculated from the incident optical power, fill factor or

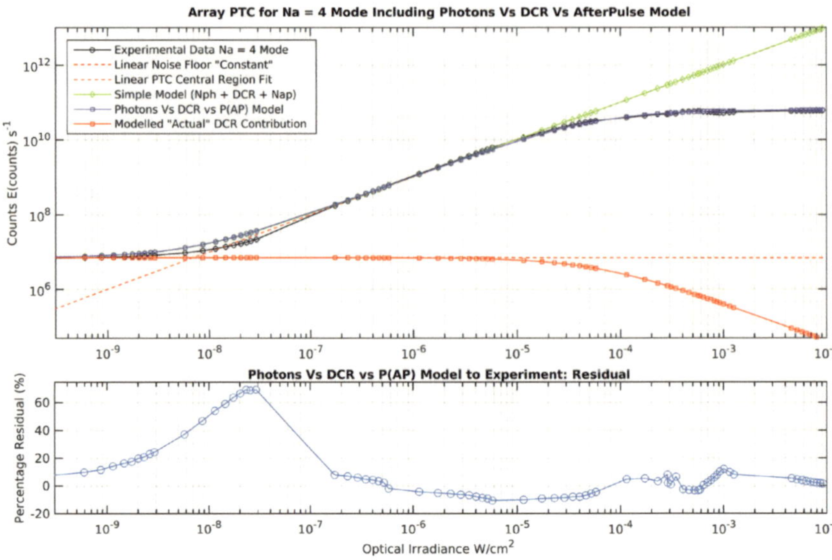

Figure 12. Count competitive and simple addition models for the photon transfer curve. The fitting parameters are: λ = 450 nm, *PDE* = 23%, *FF* = 2.4%, *P(ap)* = 1%, T_d = 16 ns, DCR = array average DCR. Two linear fits are included to both the central region of the experimental PTC and the constant noise floor. The competition implies a reduced DCR as the PTC saturates, this is also included.

carrier generation rate etc. The final counts N_{counts}, including competitive interaction and the finite dead time, is then given by Eq. (19).

$$N_{counts} = \left(\frac{\Delta T}{1+(1+P(ap))(\psi_0+DCR_0)(T_d)}\right)(\psi_0+DCR_0)+\left(\frac{\Delta T}{1+(1+P(ap))(\psi_0+DCR_0)(T_d)}\right)^2(\psi_0+DCR_0)P(ap) \quad (19)$$

In **Figure 12**, this model shows a low overall residual error, while incorporating the saturation at the level of C_{MAX}. The deviation from experimental values in the region of 1 × 10^{-8} W/cm^2 is due to small inconsistencies in the measurement approach as the optical power approached the limits of the optical power meter. The power of the model however, is that the instantaneous DCR or after-pulsing counts can be estimated given count competition rather than assuming their calculated, or stand-alone measurement values. For example, DCR_0 may be measured at a mean of 100 cps, however through the introduction of some photon detections and the corresponding loss in T_{enb}, the actual detected dark count rate, DCR_{act} may instead be 90 cps, i.e. 10 thermal generation events were ignored as they happen to coincide with SPAD dead-times caused by photon events. The value of DCR_{act} is therefore plotted showing that as a SPAD approaches count saturation, the number of detected dark events falls. This effect must therefore be included in any assessment of count noise at a particular optical power, as would be needed for the BER of OOK or M-PAM of finite extinction ratio.

4.2. Step and impulse response

The step and impulse responses play a key role in the effectiveness of communication. In [11, 12], the OOK communications experiments are severely impacted by both (i) a step response initial peak, ripple and lower than expected settled count rate and (ii) an impulse response slow decay. Both act to close the data eye, thereby increasing the BER. For the step response, the inter-symbol interference effect is directly linked to the dead-time of SPADs as a function of the array read-out-rate, and has also been observed experimentally in [39, 40].

Figure 13 shows a large change in the peak and ripple as the input power increases [11]. For the 5 μW step, the peak is 29% above the steady state. This will force a separate peak in histograms of the data amplitude levels depending on the data sequence. The following data point shows a value 10% below the steady state, indicating that a received 'one' at this point would have an artificially lowered amplitude. In comparison, lower optical powers such as the 0.5 μW step, give peaks only 10% above the steady state, followed by a minimal trough. When combined with the observation that the lower light level responses have a steady state closer to the initial, (and correct incident), number of photons, there will be a corresponding reduction in the splitting of the N_1 data distribution and an improved BER (explored further in [11]). As the steady state is below the initial number of counts, the relative decrease in steady state as optical power increases acts to decrease the received extinction ratio of the modulated OOK or M-PAM signal, thereby increasing errors.

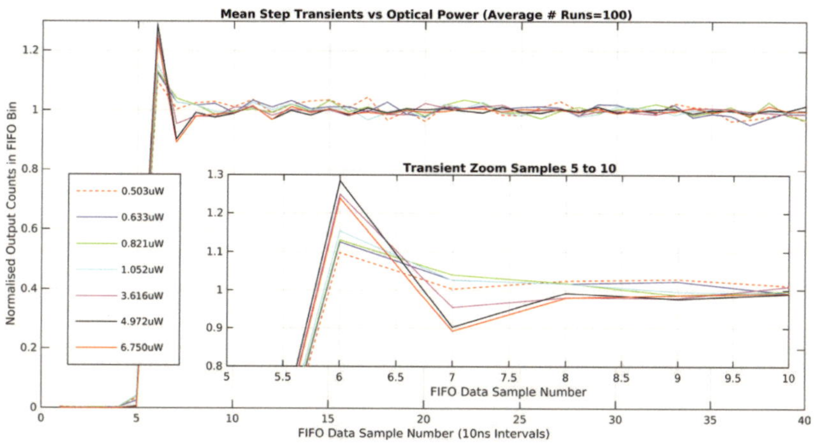

Figure 13. Normalised average step response over a sweep of optical power. The peak and ripple in the number of counts per 10 ns bin, is highlighted in the inset zoom (50 ns) after the start of the laser step.

4.2.1. A detector pooling model of SPAD array step response:

The SPAD-array step response can be modelled allowing optimisation of future designs. Initially, the receiver has a full complement of SPADs (e.g. N_{array} = 1024). At some subsequent time, a number of diodes have broken down, giving a number at the output. These have been removed from the pool, and hence the effective total efficiency of the array, (i.e. the PDE and array fill-factor for an array of N_{array} SPADs), η_{det}, for subsequent photon arrivals reduces as there is a new number of available SPADs, N_{avail}. The removal of SPADs from the pool, and the reduction in detection probability, leads to decreased output counts despite continuous input levels. Sometime later, the SPADs that were removed will be recharged back into the available pool, bringing the effective detection efficiency back towards unity.

Eq. (20) is an initial recursive discrete time model for the number of SPADs available at a time step, n, where $C_{in}(n)$ is the expected number of input counts (photons, dark counts and afterpulses). The model makes several assumptions, however fitting to experimental results shows that: (i) the assumptions hold for the 32 × 32 digital summation integration mode receiver and (ii) these can be tackled in subsequent revisions of the model. The model assumes that the SPADs break down at the start of a discrete time step, and that the array readout frequency giving rise to these discrete time steps is initially equal to the SPAD dead time. This latter assumption is expanded within the modelling below to include dead times integer multiples of the readout period, i.e. $2n$, $3n$ etc. It should be noted that the quantity, $PDE_0FF_0C_{in}(n)$ models the number of SPAD detections received by the SPAD array within the discrete period and thus can be combined with the count competitive model (Eq. (19)).

$$N_{avail}(n) = N_{avail}(n-1) - \frac{PDE_0\,FF_0\,N_{avail}(n-1)\,C_{in}(n-1)}{N_{array}} + \frac{PDE_0\,FF_0\,N_{avail}(n-2)\,C_{in}(n-2)}{N_{array}} \tag{20}$$

The steady state and the reduction of the extinction ratio due to this effect, can be estimated by Eq. (21), (the 'alpha model'), or a simple model (Eq. (22)) using the initial input counts (see [11]). The 'alpha model' uses the long-term average input counts, $\langle C_{in}\rangle$, expected for a light level, i.e. N_{counts} from Eq. (19), along with a fitting factor, α, which has been found through minimisation of the residual to be 0.725 i.e. approximately $1/\sqrt{2}$.

$$C_{out}(n) = \langle C_{in}\rangle\left[1 - \frac{\langle C_{in}\rangle}{N_{array}}\left(1 + \alpha\frac{\langle C_{in}\rangle}{N_{array}}\right)\right] \tag{21}$$

$$C_{out}(n) = \langle C_{in}\rangle\left[1 - \frac{\langle C_{in}\rangle}{N_{array}}\right] \tag{22}$$

The discrete time model (Eq. (20)) is (i) uncalibrated, (ii) departs from noisy experimental results and (iii) is not yet combined with the steady state model of Eq. (21). Despite this it can be used as an indication as to future modelling and device requirements. The critical point of both time and steady state models is that they allow preliminary investigation into the splitting of the OOK N_1 distribution that is seen clearly in experimental communication results (next section). In **Figure 14**, an array of 1024 SPADs receives a step change in optical power. The input is fixed at an example 100 photons per symbol as ultimately the BER must remain static (Eq. (5)) [1, 6]. With the dead time fixed at 10 ns, the discrete time 'readout' period is changed as a proxy for different data rates. As the data symbol or 'readout' period becomes shorter than the dead time, there is a severe increase in the ripple, the number of symbols prior to SPADs recharging into the detector pool and a decrease in the steady state. While readout periods of 20 ns for the 10 ns dead time produce minimal decreases in the effective detection efficiency and step-response ripple, we must remember that there may be a limit to how short the dead time can be. As designs push the data rate in the first instance, a dead-time equal to the readout or symbol period is preferable.

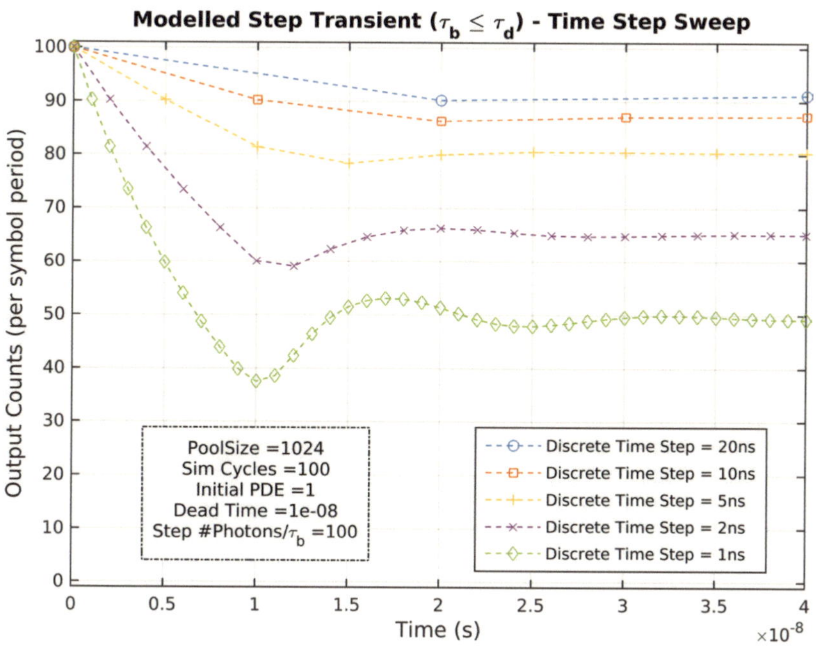

Figure 14. Modelled (uncalibrated) step response with an input step of 100 photons per symbol, a pool of 1024 SPADs and a SPAD dead time of 10 ns. The discrete time step or array readout speed is swept as a proxy for data rates between 50 Mb/s and 1 Gb/s.

In **Figure 15**, a readout period of 1 ns is chosen to highlight observations within the model. The step is held at 100 photons per symbol and the dead time is again set to 10 ns, however here the size of the array is changed from 2048 to 256 SPADs. It is possible to decrease the severity of the step response by increasing the number of diodes. However, clearly it is the ratio of the expected counts per symbol to the pool size that is critical. Reduce the number of diodes too much and the step response ripple becomes the limiting factor for both communications (BER) and optical metrology (for 100 photons per sample, the steady state of the 256 SPAD array is approx. 20 counts per sample). For completeness, the model, (once calibrated), can be used to estimate the number of diodes necessary, at a specified dead time (10 ns) and symbol period (1 ns), for a step response ripple and steady state that is within the Poisson noise of the expected input count level. For 100 counts per symbol, the standard deviation, ($\sqrt{100} = 10$), leads to ~9750 SPADs required for a steady state of 90.

These parameter sweeps, although chosen to visually highlight SPAD array issues, point to three heuristic design rules:

- First, that the dead-time should be equal to the data rate, (10 ns for 100 Mb/s etc.),
- Second, that the number of SPADs should be high in comparison to the expected photon counts per data symbol, and
- Third, if T_d is limited, the number of diodes can be increased to reduce the detrimental effects on the step response if we wish to operate at a data rate higher than $1/T_d$.

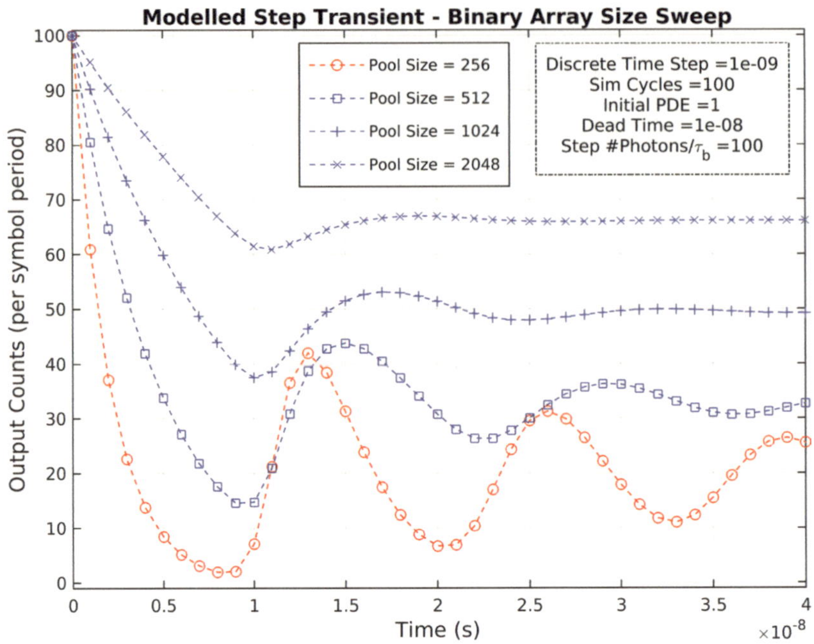

Figure 15. Modelled (uncalibrated) step response with an input step of 100 photons per symbol, a SPAD dead time of 10 ns and an array readout period of 1 ns (for illustration purposes). The number of SPADs in the pool is swept highlighting that if a dead time equal to the symbol period is not possible, the number of SPADs must be large in comparison to the expected photon arrivals per symbol.

Of course, these rules are a product of only the step response model. Increasing the number of diodes within the array may have a detrimental effect on the array fill factor. Hence research must concentrate on models that once combined, can be used for overall receiver optimisation with respect to communication level metrics (sensitivity, BER, data rate, etc.).

4.3. Communication results and feasibility

Early communications results have been achieved using the 32 × 32 SPAD-based integrating receiver, demonstrating that such receivers are feasible. **Figure 16** shows the eye diagram of a 50 Mb/s on-off key non-return-to-zero data stream. The transmitter in this case is a Gallium-Nitride 34 μm diameter 450 nm micro-LED, with a bandwidth of 220 MHz. The data eye is open at this optical power (20 μW/cm^2) and data rate, however several issues become apparent.

- Firstly, upon a 'zero' to 'one' transition, there is a large peak 75–100 counts higher than the main N_1 level (approx. 250). This can be directly attributed to the step response discussed above, and can be seen clearly near sample 110 in the time traces of **Figure 17**. Indeed, it leads to a split N_1 distribution, with the severity of the splitting of the ideally two distributions (N_0 and N_1) being evident in the count histogram.
- Secondly, the impulse response, (measured in [11]), creates a slow 'one' to 'zero' transition. This is clearly visible near time sample 90 (**Figure 17**). For some high frequency transitions, the impulse response prevents the transmitted 'zero' from obtaining the low count average

(approx. 45) as the impulse response is still decaying when the next 'one' bit is transmitted (see near sample 150). This therefore leads to the splitting of the N_0 distribution.

While excellent agreement is shown between the transmitted and received pseudo random bit streams (**Figure 17**), the closure of the eye can be attributed to (a) the steady state of the step response, (b) the bleeding of previous 'one' bits into subsequent 'zero' bits due to the impulse response, and (c) the finite extinction ratio of the micro-LED transmitter. The 32 × 32 SPAD receiver, as a testing harness, has provided crucial details for the future development of SPAD-based receivers in that clearly, the transient response which

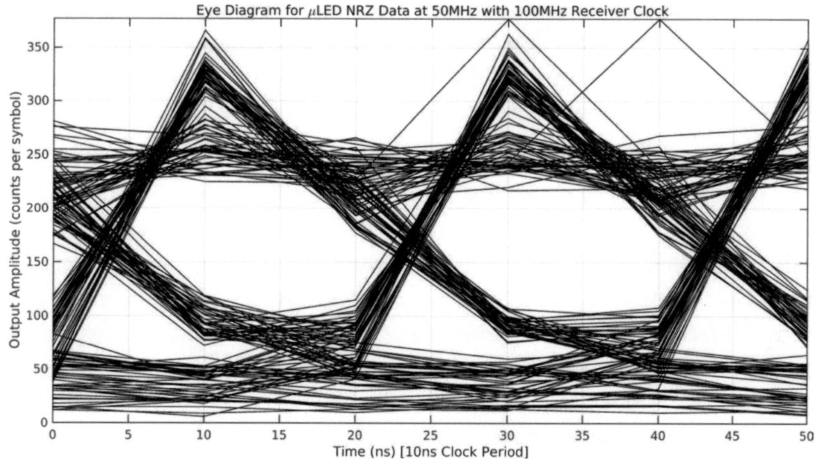

Figure 16. A measured eye diagram for 50 Mb/s OOK-NRZ data. The dead time was approx. 13.5 ns.

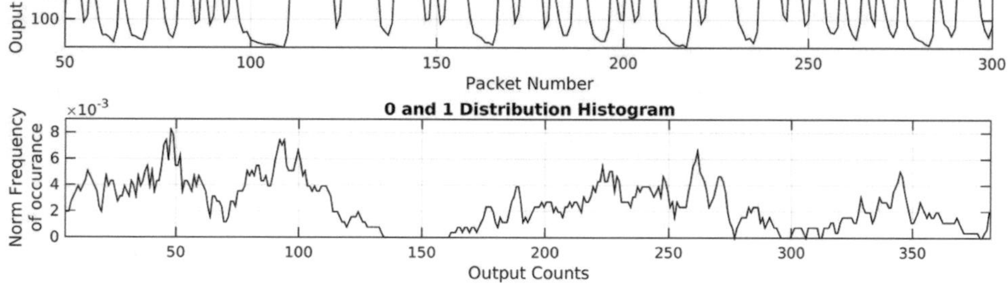

Figure 17. The transmitted OOK NRZ data stream, a time trace of the SPAD-receiver output and the histogram of received amplitudes. A sampling threshold at 150 counts shows separability of the N_0 and N_1 distributions, however the impulse and step responses have acted to split the distributions and close the data eye. With a reduced impulse and step response, the transmitter should render Poisson-limited distributions near 50 and 350 counts.

was only evident once empirically measured, must be addressed for low bit error rate communications.

5. Conclusions

SPAD-based optical receivers increase the prospective methodologies for applications requiring extremely high optical sensitivity. This chapter has covered the specifics of both optical detection using the internal photoelectric effect, and how avalanche multiplication can be extended to run-away avalanche producing full logic-level voltage pulses with single-photon sensitivity. The application, visible light communications, highlights that such receivers require further study before all noise sources and all circuit topologies are understood fully with respect to the higher-level communication performance metrics.

Three SPAD receiver topologies were introduced, the digital summation, the XOR summation and a current-summation. The performance metrics of both (i) optical communications (data rate, bit error rate, energy per bit etc.) and (ii) single SPAD and SPAD array circuits (fill factor, PDE, DCR), were also discussed. At present, there is little receiver theory to link low-level SPAD metrics with higher-level performance, although some work as already been done. For example, both the SPAD array fill factor and photon detection efficiency are known to directly impact optical sensitivity, and appropriate modelling has been developed. The receiver BER has also been investigated with respect to the SPAD dead time and the output photon-countingstatistics of a digital summation SPAD-receiver. To date however, there is little to link the SPAD after-pulsing probability with the BER, although one would assume its inherent delay would act as a form of ISI.

One requirement for future research and commercialization, would be the use of application specific design methodologies fully linked to higher level communications performance. How should the designer balance the SPAD fill factor and dead-time for example, when attempting to optimize a receiver for a required data rate and sensitivity? At present, too few high to low performance relations exist to allow such an interplay to be optimized. Likewise, if the application speed requires the dead-time to be traded off with after-pulsing, the effect that after-pulsing has needs to be quantified to allow comparison with dead-time ISI effects.

Overall, SPAD-based receivers are an active field of research, aiming to provide optimized designs with both extreme sensitivity and high-speed. Data-rates and energy per bit will likely remain modest and far removed the highest speed links of today. However, such receivers may offer options in (i) room-lighting VLC applications, supplementing existing Wi-Fi for download intensive tasks such as high-quality, real-time video streaming, and (ii) niche applications where optical attenuation is significant enough to warrant increased receiver gain and a higher energy per bit.

6. Student exercise questions

1. Calculate the quantum limit number of photons required per symbol, assuming an ideal, noiseless receiver, at a data rate of 50 Mb/s, a BER of 1×10^{-6}, and a wavelength of 600 nm.

2. Calculate the minimum number of SPADs within a receiver array, for the conditions in question #1, if all SPADs have a dead time of 10 ns.

3. If the SPADs are circular, and laid out in a square array, calculate the approximate array fill factor (10 μm SPAD diameter), assuming (i) no CMOS circuitry is incorporated within the array and (ii) that D_s and D_g are 2 μm and 5 μm respectively.

Author details

Edward M.D. Fisher

Address all correspondence to: e.fisher@ed.ac.uk

Institute of Digital Communications (IDCOM), The School of Engineering, The University of Edinburgh, Scotland, UK

References

[1] Razavi B. Design of Integrated Circuits for Optical Communications. 1st ed. McGraw-Hill, New York, USA; 2003.

[2] Binh LN. Digital Optical Communications. 1st ed. CRC Press, Boca Raton, Florida, USA; 2008.

[3] Hanzo L, Haas H, Imre S, O'Brien D, Rupp M, Gyongyosi Y. Wireless myths, realities, and futures: From 3G/4G to optical and quantum wireless. Proceedings of the IEEE. 2012;**100**:1853-1888 (Special Centennial Issue)

[4] Elgala H, Mesleh R, Haas H. Indoor broadcasting via white LEDs and OFDM. IEEE Transactions on Consumer Electronics. 2009;**55**(3):1127-1134.

[5] Cisco Systems Inc. Cisco Visual Networking Index: Global Mobile Data Traffic Forecast Update, 2016-2021 White Paper [Internet]. 09/02/17 [Updated: 09/02/17]. Available from: http://www.cisco.com/c/en/us/solutions/collateral/service-provider/visual-networking-index-vni/mobile-white-paper-c11-520862.pdf [Accessed: 24/02/17]

[6] Sackinger E. Broadband Circuits for Optical Fiber Communication. Wiley, New Jersey, USA; 2005.

[7] Carusone A, Yasotharan H, Kao T. Progress and trends in multi-Gbps optical receivers with CMOS integrated photodetectors. In: Custom Integrated Circuits Conference (CICC); IEEE, San Jose California, USA; 2010:1-8.

[8] Zeng L, O'Brien D, Minh H, Faulkner G, Lee K, Jung D, Oh Y, Won ET. High data rate multiple input multiple output (MIMO) optical wireless communications using white LED lighting. IEEE Journal on Selected Areas in Communications. 2009;**27**(9):1654-1662.

[9] O'Brien D, Le Minh H, Zeng L, Faulkner G, Lee K, Jung D, Oh Y, Won ET. Indoor visible light communications: challenges and prospects: free-space laser communications VIII. Proceedings of the SPIE. 2008;**7091**:709106-709109.

[10] IEEE Computer Society—WG802.15—Wireless Personal Area Network (WPAN) Working Group. P802.15.7 —IEEE Approved Draft Standard for Short Range Wireless Optical Communication Using Visible Light [Internet]. April 2011 [Updated: Rev. P802.15.7]. Available from: https://standards.ieee.org/develop/project/802.15.7.html

[11] Fisher E. A parallel, reconfigurable single-photon avalanche diode array for optical communications [Thesis]. The School of Engineering, The University of Edinburgh, Edinburgh, Scotland, UK; 2015.

[12] Fisher E, Underwood I, Henderson R. A reconfigurable single-photon-counting integrating receiver for optical communications. IEEE Journal of Solid-State Circuits. 2013;**48**(7): 1638-1650.

[13] Henley E, Garcia A. Sub-Atomic Physics. 3rd ed. World Scientific, Toh Tuck Link, Singapore; 2007.

[14] Durini D, editor. High Performance Silicon Imaging: Fundamentals and Applications of CMOS and CCD Sensors. 1st ed. Elsevier Ltd, Woodhead Publishing (for Elsevier), Cambridge, UK; 2014.

[15] Roychoudhuri C, Kracklauer AF, Creath K, editors. The Nature of Light: What is a Photon?. 1st ed. CRC Press, Boca Raton, Florida, USA; July 2008. 452 p.

[16] Sze SM. Semiconductor Devices: Physics and Technology. 2nd ed. Wiley and Sons, New Jersey, USA; 2001.

[17] Campbell J. Recent advances in telecommunications avalanche photodiodes. Journal of Lightwave Technology. 2007;**25**(1):109-121.

[18] Yang B, Schaub J, Csutak S, Campbell J. 10 Gbps all silicon APD optical receiver. Lasers and Electro-Optics Society. The 15th Annual Meeting of the IEEE. 2002;**2**:681-682.

[19] Takahashi S, Tajima A, Tomita A. High-efficiency single photon detector combined with an ultra-small APD module and a self-training discriminator for high-speed quantum cryptosystems. In: Proc. 13th Micro-Optics Conference, Japan. 2007

[20] Ghioni M, Gulinatti A, Rech I, Zappa F, Cova S. Progress in silicon single-photon avalanche diodes. IEEE Journal of Selected Topics in Quantum Electronics. 2007;**13**(2):852-862.

[21] Haitz RH. Mechanisms contributing to the noise pulse rate of avalanche diodes. Journal of Applied Physics. 1965;**36**(10):3123-3131.

[22] Webster E. Single-photon avalanche diode theory, simulation, and high performance CMOS integration [Thesis]. The School of Engineering: University of Edinburgh; 2013.

[23] Cova S, Ghioni M, Lacaita A, Samori C, Zappa F. Avalanche photodiodes and quenching circuits for single-photon detection. Applied Optics. 1996;35(12):1956-1976.

[24] Richardson J, Webster E, Grant L, Henderson R. Scaleable single-photon avalanche diode structures in nanometer CMOS technology. IEEE Transactions on Electron Devices. 2011;**58**(7):2028 -2035.

[25] Goetzberger A, McDonald B, Haitz R, Scarlett RM. Avalanche effects in silicon p-n junctions. II. Structurally perfect junctions. Journal of Applied Physics. 1963;**34**(6):1591-1600.

[26] Richardson J, Grant L, Henderson R. Low dark count single-photon avalanche diode structure compatible with standard nanometer scale CMOS technology. IEEE Photonics Technology Letters. 2009;**21**(14):1020 -1022.

[27] Webster E, Richardson J, Grant L, Henderson R. Single-photon avalanche diodes in 90 nm CMOS imaging technology with sub-1 Hz median dark count rate. Proceedings of the International Image Sensor Workshop. 2011:262-265.

[28] Lee S, Gardner R. A new G-M counter dead time model. Applied Radiation and Isotopes. 2000;**53**(4-5):731-737.

[29] Neri L, Tudisco S, Musumeci F, Scordino A, Fallica G, Mazzillo M, Zimbone M. Note: dead time causes and correction method for single photon avalanche diode devices. Review of Scientific Instruments. 2010;**81**(8): 086102-1 to 086102-3

[30] Blazej J, Prochazka I. Avalanche photodiode output pulse rise-time study. In: SPIE 7355, Photon Counting Applications, Quantum Optics, and Quantum Information Transfer and Processing II; May; SPIE Headquarters: Bellingham, Washington, USA; 2009. pp. 73550M-11.

[31] Spinelli A, Lacaita A. Physics and numerical simulation of single-photon avalanche diodes. IEEE Transactions on Electron Devices. 1997;**44**(11):1931-1943.

[32] Vilella E, Alonso O, Montiel A, Vila A, Dieguez A. A low-noise time-gated single-photon detector in a HV-CMOS technology for triggered imaging. Sensors and Actuators A: Physical. 2013;**201**(1):342-351.

[33] C. Niclass, M. Soga, H. Matsubara and S. Kato, "A 100m-range 10-frame/s 340×96-pixel time-of-flight depth sensor in 0.18μm CMOS," 2011 Proceedings of the ESSCIRC (ESSCIRC), Helsinki, 2011, pp. 107-110. doi: 10.1109/ESSCIRC.2011.6044926

[34] Rech I, Ingargiola A, Spinelli R, Labanca I, Marangoni S, Ghioni M, Cova S. Optical cross-talk in single photon avalanche diode arrays: a new complete model. Optics Express. 2008;**16**(12):8381-8394.

[35] Zhang G, Yu C, Zhu C, Liu L. Feasibility study of multi-pixel photon counter serving as the detector in digital optical communications. Elsevier Optik. 2013;**124**(22):5781-5786. DOI: http://dx.doi.org/10.1016/j.ijleo.2013.04.060

[36] Dutton N, Gnecchi S, Parmesan L, Holmes A, Rae B, Grant L, Henderson R. A Time-correlated single-photon-counting sensor with 14GS/s histogramming time-to-digital converter. In: Technical Digest International Solid-State Circuits Conference (ISSCC); 22 Feb; San Francisco. All IEEE references share the same IEEE headquarters location of: New York, USA; 2015. pp. 1-3.

[37] Gnecchi S, Dutton N, Parmesan L, Rae B, Pellegrini S, McLeod S, Grant L, Henderson R. Digital silicon photomultipliers with OR/XOR pulse combining techniques. IEEE Transactions on Electron Devices. 2016;**63**(3):1105-1110.

[38] Almer O, Dutton N, Abbas T, Gnecchi S, Henderson R. 4-PAM visible light communications with a XOR-tree digital silicon photomultiplier. In: Summer Topicals Meeting Series (SUM); 13 July; IEEE; 2015. pp. 41-42.

[39] Chitnis D, Collins S. A SPAD-based photon detecting system for optical communications. Journal of Lightwave Technology. 2014;**32**(10):2028-2034.

[40] Chitnis D, Zhang L, Chun H, Rajbhandari S, Faulkner G, O'Brien D, Collins S. A 200Mb/s VLC demonstration with a SPAD based receiver. In: Summer Topicals Meeting Series (SUM), 2015; July; All IEEE Publisher Locations are New York, USA; 2015. p. 226-227.